Lecture Notes in Statistics 178

Edited by P. Bickel, P. Diggle, S. Fienberg, K. Krickeberg,
I. Olkin, N. Wermuth, and S. Zeger

Springer
New York
Berlin
Heidelberg
Hong Kong
London
Milan
Paris
Tokyo

Gauri Sankar Datta
Rahul Mukerjee

Probability Matching Priors: Higher Order Asymptotics

 Springer

Gauri Sankar Datta
Department of Statistics
University of Georgia
Athens, GA 30602-1952
USA
gauri@stat.uga.edu

Rahul Mukerjee
Indian Institute of Management
Joka, Diamond Harbour Road
Post Box No. 16757
Alipore Post Office
Calcutta 700 027
India
rmuk1@hotmail.com

Library of Congress Cataloging-in-Publication Data
Datta, Gauri Sankar.
    Probability matching priors : higher order asymptotics / Gauri Sankar Datta, Rahul Mukerjee.
      p.   cm — (Lecture notes in statistics ; 178)
    Includes bibliographical references and index.

      1. Mathematical statistics—Asymptotic theory.   2. Bayesian statistical decision theory.   3.
    Probabilities.   I. Mukerjee, Rahul.   II. Title.   III. Lecture notes in statistics
    (Springer-Verlag) ; v. 178.
    QA276 .D32555
    519.5—dc22   2003066221

ISBN-13: 978-0-387-20329-4   e-ISBN-13: 978-1-4612-2036-7

DOI: 10.1007/978-1-4612-2036-7

Printed on acid-free paper.

9 8 7 6 5 4 3 2 1   SPIN 10937969

Springer-Verlag is a part of *Springer Science+Business Media*

springeronline.com

Lecture Notes Editorial Policies

Lecture Notes in Statistics provides a format for the informal and quick publication of monographs, case studies, and workshops of theoretical or applied importance. Thus, in some instances, proofs may be merely outlined and results presented which will later be published in a different form.

Publication of the Lecture Notes is intended as a service to the international statistical community, in that a commercial publisher, Springer-Verlag, can provide efficient distribution of documents that would otherwise have a restricted readership. Once published and copyrighted, they can be documented and discussed in the scientific literature.

Lecture Notes are reprinted photographically from the copy delivered in camera-ready form by the author or editor. Springer-Verlag provides technical instructions for the preparation of manuscripts. Volumes should be no less than 100 pages and preferably no more than 400 pages. A subject index is expected for authored but not edited volumes. Proposals for volumes should be sent to one of the series editors or addressed to "Statistics Editor" at Springer-Verlag in New York.

Authors of monographs receive 50 free copies of their book. Editors receive 50 free copies and are responsible for distributing them to contributors. Authors, editors, and contributors may purchase additional copies at the publisher's discount. No reprints of individual contributions will be supplied and no royalties are paid on Lecture Notes volumes. Springer-Verlag secures the copyright for each volume.

Series Editors:

Professor P. Bickel
Department of Statistics
University of California
Berkeley, California 94720
USA

Professor P. Diggle
Department of Mathematics
Lancaster University
Lancaster LA1 4YL
England

Professor S. Fienberg
Department of Statistics
Carnegie Mellon University
Pittsburgh, Pennsylvania 15213
USA

Professor K. Krickeberg
3 Rue de L'Estrapade
75005 Paris
France

Professor I. Olkin
Department of Statistics
Stanford University
Stanford, California 94305
USA

Professor N. Wermuth
Department of Psychology
Johannes Gutenberg University
Postfach 3980
D-6500 Mainz
Germany

Professor S. Zeger
Department of Biostatistics
The Johns Hopkins University
615 N. Wolfe Street
Baltimore, Maryland 21205-2103
USA

To my departed mother and to my wife, GSD
To my parents and wife, RM

Acknowledgements

We record our deep gratitude to Professor J.K. Ghosh for initiating us to the fascinating area of probability matching priors, and also for the guidance and support that he had always extended to us.

We very much appreciate the encouragement and support that we received from Professors Malay Ghosh, Nancy Reid and C.F. Jeff Wu in this endeavor. We thank two anonymous referees for their very constructive suggestions.

G.S. Datta was supported by Professor James O. Berger for a visit to Purdue University in the fall of 1993. During this visit Datta had an excellent exposure to the topic of this monograph from Professor J.K. Ghosh. For this, he is deeply indebted to both of them.

We acknowledge the assistance received for this project from the University of Georgia, and the Center for Management and Development Studies, Indian Institute of Management Calcutta.

Finally, we are grateful to our families for their encouragement and support in various ways.

August, 2003, Athens, GA, USA   *Gauri S. Datta*
Kolkata, India   *Rahul Mukerjee*

Contents

Contents

1

Introduction and the Shrinkage Argument

1.1 Scope of the monograph

The study of priors ensuring, up to the desired order of asymptotics, the approximate frequentist validity of posterior credible sets has received significant attention in recent years and a considerable interest is still continuing in this field. Bayesian credible sets based on these priors have approximately correct frequentist coverage as well. Such priors are generically known as *probability matching priors*, or *matching priors* in short. As noted by Tibshirani (1989) among others, study in this direction has several important practical implications with appeal to both Bayesians and frequentists:

(a) First, the ensuing matching priors are, in a sense, noninformative. The approximate agreement between the Bayesian and frequentist coverage probabilities of the associated credible sets provides an external validation for these priors. They can form the basis of an objective Bayesian analysis and are potentially useful for comparative purposes in subjective Bayesian analyses as well.

(b) Second, Bayesian credible sets given by matching priors can also be interpreted as accurate frequentist confidence sets because of their approximately correct frequentist coverage. Thus the exploration of matching priors provides a route for obtaining accurate frequentist confidence sets which are meaningful also to a Bayesian.

(c) In addition, research in this area has led to the development of a powerful and transparent Bayesian route, via a shrinkage argument, for higher order asymptotic frequentist computations.

The present monograph aims at reviewing the recent developments in and around matching priors. While we discuss or refer to topics that, in our opinion, are most important and useful, the monograph is not intended to be encyclopedic. Reference is made to Ghosh and Mukerjee (1998) for an earlier review which, however, does not contain any proof. We also refer to Ghosh and

Mukerjee (1992a), Reid (1995) and Kass and Wasserman (1996) for discussion on matching priors among other things. A special feature of the present monograph is that we have attempted to show the the key steps leading to the main results in some detail that is not available in the existing literature which is scattered in journals and proceedings. The regularity conditions that justify these steps have been informally discussed (see Section 2.2). For the interested reader, detailed references for these conditions have also been given. It is hoped that this approach will make the monograph accessible to a wider audience including non-specialists.

The shrinkage argument mentioned in (c) above is a central tool in the characterization of matching priors. In fact, it can be useful also in purely frequentist problems by significantly reducing the underlying algebra. Therefore, in the rest of this chapter, we describe the shrinkage argument and illustrate it with an example.

The subsequent chapters of the monograph are organized as follows. In Chapter 2, we review matching priors that ensure approximate frequentist validity of posterior quantiles when the interest parameter is one-dimensional. Matching priors of this kind are arguably the most widely studied ones in the literature. With multiple parameters or parametric functions of interest, however, posterior quantiles are not well-defined and consideration of the posterior cumulative distribution function can be of help. This issue has been taken up in Chapter 3.

In Chapter 4, we study priors ensuring approximate frequentist validity of highest posterior density (HPD) regions which have been very popular with Bayesians as "smallest" credible sets. These results can also be of particular use in getting matching priors when the interest parameter is multidimensional. Chapter 5 deals with similar problems on the basis of other posterior credible sets obtained via the inversion of commonly used statistics including the likelihood ratio statistic. Here we further discuss how for a given prior the HPD or ellipsoidal regions can be perturbed so as to achieve both Bayesian and frequentist validity.

While the development up to Chapter 5 depends on the specification of an interest parameter which may be one-dimensional or multidimensional, in Chapter 6 we consider the situation where one is concerned with the prediction of a future observation rather than the estimation of an interest parameter. Priors ensuring approximate frequentist validity of posterior prediction regions are discussed here. Chapter 6 and also the monograph end with some brief remarks on areas that deserve further attention.

The interconnection among the results in various chapters has been explored whenever possible.

Before concluding this section, in continuation of (a) above, we note that there are other approaches too for the development of noninformative priors, the most notable of these approaches being the one based on reference priors (Bernardo, 1979; Berger and Bernardo, 1992a). A reference prior is operationally defined as a prior which maximizes some appropriate missing information about the interest parameter, measured through a suitable information-theoretic functional (Bernardo and Smith, 1994, Ch. 5). A related approach, based on information tradeoff priors (Clarke and Wasserman, 1995), has also received attention. In this monograph, however, we concentrate on matching priors. Only in examples, occasionally we make casual mention of reference priors. Highly illuminating accounts of reference priors are already available in Bernardo and Smith (1994) and Bernardo and Ramon (1998). In addition, a catalog of reference and several other noninformative priors appears in Yang and Berger (1997). Interestingly, in quite a few of our examples, reference priors turn out to be probability matching as well. When this is not the case, we believe that a choice between the two approaches is a matter of taste.

1.2 The shrinkage argument

The shrinkage argument, which is a crucial technique in the development of matching priors, was suggested originally by J.K. Ghosh to the present authors (Ghosh and Mukerjee, 1991; Ghosh, 1994, Ch. 9). It is foreshadowed by Bickel and Ghosh (1990) and is reminiscent of Dawid (1991). Accounts of the shrinkage argument are available in Mukerjee and Reid (2000) and the unpublished thesis of Li (1998). In this and the next section, we closely follow Mukerjee and Reid (2000).

Consider a possibly vector-valued random variable X with a probability density function $g(\cdot; \theta)$ where the parameter θ belongs to the $p-$dimensional Euclidean space $\mathcal{R}^p$ or some open subset thereof. It is intended to find an expression for the expectation $E_\theta\{q(X, \theta)\}$, where q is a measurable function. The expectation is known to exist and is supposed to be continuous for all θ. Of special interest in the present context is the situation where q is an indicator function in which case $E_\theta\{q(X, \theta)\}$ represents a frequentist probability. The following steps describe a Bayesian approach for the evaluation of $E_\theta\{q(X, \theta)\}$.

Step 1: Consider a proper prior density $\bar{\pi}(\cdot)$ for θ, such that the support of $\bar{\pi}(\cdot)$ is a compact rectangle in the parameter space and $\bar{\pi}(\cdot)$ vanishes on the boundary of the support while remaining positive in the interior. As usual, the support of $\bar{\pi}(\cdot)$ is the closure of the set on which it is positive. Consider the posterior density of θ under $\bar{\pi}(\cdot)$, and hence obtain $E^{\bar{\pi}}\{q(X, \theta)|X\}$, which is the expectation of $q(X, \theta)$ in the posterior setup.

Step 2: Find $E_\theta E^{\bar{\pi}}\{q(X, \theta)|X\}(= \lambda(\theta), \text{say})$, for θ in the interior of the support of $\bar{\pi}(\cdot)$.

Step 3: Integrate $\lambda(\cdot)$ with respect to $\bar{\pi}(\cdot)$ and then allow $\bar{\pi}(\cdot)$ to converge weakly to the degenerate prior at the true θ, supposing that the true θ is an interior point of the support of $\bar{\pi}(\cdot)$. This yields $E_\theta\{q(X, \theta)\}$.

We now indicate the rationale behind the above steps which assume the integrability of $q(X, \theta)$ with respect to the joint probability measure for (X, θ) as induced by $\bar{\pi}(\cdot)$. Such integrability allows the changes in the order of integration in what follows and holds, in particular, if q is an indicator function.

Note that the posterior density of θ under the prior $\bar{\pi}(\cdot)$ is given by $g(X;\theta)\bar{\pi}(\theta)/N(X)$, where

$$N(X) = \int g(X;\theta)\bar{\pi}(\theta)\mathrm{d}\theta . \tag{1.2.1}$$

Hence Step 1 yields

$$E^{\bar{\pi}}\{q(X,\theta)|X\} = K(X)/N(X) , \tag{1.2.2}$$

where

$$K(X) = \int q(X,\theta)g(X;\theta)\bar{\pi}(\theta)\mathrm{d}\theta. \tag{1.2.3}$$

By (1.2.2), Step 2 yields

$$\lambda(\theta) = \int \{K(x)/N(x)\}g(x;\theta)\mathrm{d}x .$$

Hence in Step 3, integrating $\lambda(\cdot)$ with respect to $\bar{\pi}(\cdot)$, we get

$$\begin{aligned}
\int \lambda(\theta)\bar{\pi}(\theta)\mathrm{d}\theta &= \int \int \{K(x)/N(x)\}g(x;\theta)\bar{\pi}(\theta)\mathrm{d}x\mathrm{d}\theta \\
&= \int \{K(x)/N(x)\}\{\int g(x;\theta)\bar{\pi}(\theta)\mathrm{d}\theta\}\mathrm{d}x \\
&= \int K(x)\mathrm{d}x \\
&= \int \int q(x,\theta)g(x;\theta)\bar{\pi}(\theta)\mathrm{d}\theta\mathrm{d}x \\
&= \int \{\int q(x,\theta)g(x;\theta)\mathrm{d}x\}\bar{\pi}(\theta)\mathrm{d}\theta \\
&= \int [E_\theta\{q(X,\theta)\}]\bar{\pi}(\theta)\mathrm{d}\theta . \tag{1.2.4}
\end{aligned}$$

In the above string of equalities, the third and the fourth equalities follow from (1.2.1) and (1.2.3), respectively. In view of the assumed continuity of $E_\theta\{q(X,\theta)\}$ for all θ and the compactness of the support of $\bar{\pi}(\cdot)$, the truth of the claim made in Step 3 follows from the last line of (1.2.4).

Note that in Step 3, $\bar{\pi}(\cdot)$ is allowed to converge to a degenerate prior. Because of this reason, the present Bayesian approach is said to be based on a *shrinkage* argument. As illustrated in the next section, explicit specification of the prior $\bar{\pi}(\cdot)$ is not needed in applying Steps 1–3 above. When executed up to the desired order of approximation under suitable assumptions, these steps can significantly reduce the algebra underlying frequentist higher order asymptotic computations. This simplification arises from two counts. First, although the Bayesian approach to frequentist asymptotics requires Edgeworth and other assumptions, it avoids an explicit calculation of Edgeworth expansion and the associated, often cumbersome, derivation of approximate cumulants. Second, it helps in expressing the final result in terms of derivatives (cf. (1.3.14) below) and hence in a compact form which is often easy to interpret. The example in the next section illustrates these points.

The shrinkage argument will be used extensively in the present monograph. In addition to substantially simplifying the derivation of matching priors in various contexts, it is also of great help in purely frequentist problems such as the one in the next section. A further illustration of the latter point will be discussed in Chapter 5 where we consider the frequentist Bartlett adjustment. We refer to Ghosh and Mukerjee (1994a) for yet another application of the shrinkage argument in the study of higher order power of tests.

1.3 An example

Let $X_1, X_2, \ldots, X_n$ be independently and identically distributed (i.i.d.) possibly vector-valued random variables with a common density $f(x; \theta)$ where the parameter θ is one-dimensional. The parameter space for θ is $\mathcal{R}^1$ or some open subset thereof. Write $X = (X_1, \ldots, X_n)^T$ and let $\widehat{\theta}$ be the maximum likelihood estimator of θ on the basis of X. Here and throughout, the superscript T stands for transpose of a vector or a matrix. Consider the problem of evaluating the frequentist tail probability $P_\theta\{n^{1/2}(\widehat{\theta} - \theta) \leq u\}$, with margin of error $o(n^{-1/2})$, where $P_\theta\{\cdot\}$ is the frequentist probability measure under θ, and the constant u is free from n and θ.

The model assumptions for the posterior and frequentist expansions considered in this section are as in Johnson (1970, pp. 852–853, with $K = 1$ in his notation) and Bhattacharya and Ghosh (1978, p. 439, with $s = 3$ in their notation) respectively. The latter assumptions entail the existence of a valid Edgeworth expansion, with margin of error $o(n^{-1/2})$, for the distribution of $n^{1/2}(\widehat{\theta} - \theta)$. These two sets of assumptions hold under wide generality and will be discussed in Section 2.2. In particular, as a common feature of both sets, the per observation Fisher information, namely I as defined in Step 2 below, is supposed to be positive for every θ.

Let $\ell(\theta) = n^{-1} \sum_{i=1}^{n} \log f(X_i; \theta)$, and define

$$c = -\ell''(\widehat{\theta}), \quad a = \ell'''(\widehat{\theta}). \tag{1.3.1}$$

In (1.3.1) and also (1.3.6) below, the primes denote differentiation with respect to θ. The quantity c is called per observation "observed" information.

Consider a proper prior density $\overline{\pi}(\cdot)$ for θ, such that the support of $\overline{\pi}(\cdot)$ is a compact interval in the parameter space and $\overline{\pi}(\cdot)$ vanishes on the boundary of the support while remaining positive in the interior. It is assumed that $\overline{\pi}(\cdot)$ satisfies the conditions in Bickel and Ghosh (1990, Section 2). These conditions ensure that $\overline{\pi}(\cdot)$ is sufficiently smooth and control $\log \overline{\pi}(\cdot)$ and its derivatives near the boundary of the support. All formal expansions for the posterior, as used in this section, are valid for sample points in a set $\overline{S}$ over which $\widehat{\theta}$ is well-defined, $\pi(\widehat{\theta}) > 0$ and $c > 0$. An explicit but somewhat technical description of $\overline{S}$ can be given along the line of Bickel and Ghosh (1990, Section 3). The set $\overline{S}$ has P_θ−probability $1 + o(n^{-1/2})$ uniformly over compact θ−sets in the interior of the support of $\pi(\cdot)$. The implication of such uniformity will be indicated later in this section.

Observe that

$$P_\theta\{n^{1/2}(\widehat{\theta} - \theta) \le u\} = E_\theta\{q(X, \theta)\}, \qquad (1.3.2)$$

where the indicator function $q(X, \theta)$ equals unity if $n^{1/2}(\widehat{\theta} - \theta) \le u$ and zero otherwise. We now execute Steps 1–3 of Section 1.2 up to the desired order of approximation.

Step 1: The posterior density of θ under the prior $\pi(\cdot)$ is given by

$$\frac{\pi(\theta)\exp\{n\ell(\theta)\}}{\int \pi(\theta)\exp\{n\ell(\theta)\}\mathrm{d}\theta} = \frac{\pi(\theta)\exp[n\{\ell(\theta) - \ell(\widehat{\theta})\}]}{\int \pi(\theta)\exp[n\{\ell(\theta) - \ell(\widehat{\theta})\}]\mathrm{d}\theta}.$$

Hence the posterior density of $y = (nc)^{1/2}(\theta - \widehat{\theta})$ under the prior $\pi(\cdot)$ is given by, say,

$$\pi^*(y|X) = \frac{b(y, X)}{\int b(y, X)\mathrm{d}y}, \qquad (1.3.3)$$

where

$$b(y, X) = \pi\{\widehat{\theta} + (nc)^{-1/2}y\}\exp(n[\ell\{\widehat{\theta} + (nc)^{-1/2}y\} - \ell(\widehat{\theta})]). \qquad (1.3.4)$$

Since $\ell'(\widehat{\theta}) = 0$, by (1.3.1) and (1.3.4), a Taylor's expansion about $\widehat{\theta}$ yields

$$b(y, X) = \pi(\widehat{\theta})\{1 + n^{-1/2}(A_1 y + A_3 y^3)\}\exp\left(-\frac{1}{2}y^2\right) + o(n^{-1/2}), \qquad (1.3.5)$$

where

$$A_1 = c^{-1/2}\pi'(\widehat{\theta})/\pi(\widehat{\theta}), \qquad A_3 = \frac{1}{6}c^{-3/2}a. \qquad (1.3.6)$$

Hence

$$\int b(y, X)\mathrm{d}y = \pi(\widehat{\theta})\sqrt{2\pi} + o(n^{-1/2}), \qquad (1.3.7)$$

the π appearing in the right-hand side of (1.3.7) being the usual transcendental number (it should not be confused with the prior $\pi(\cdot)$). By (1.3.3), (1.3.5) and (1.3.7),

$$\begin{aligned}\pi^*(y|X) &= \phi(y)\{1 + n^{-1/2}(A_1 y + A_3 y^3)\} + o(n^{-1/2})\\ &= \phi(y)[1 + n^{-1/2}\{G_1 J_1(y) + G_3 J_3(y)\}] + o(n^{-1/2}), \qquad (1.3.8)\end{aligned}$$

where

$$G_1 = A_1 + 3A_3, \qquad G_3 = A_3, \qquad (1.3.9)$$

$\phi(\cdot)$ is the standard univariate normal density, and $J_i(\cdot)$ is the Hermite polynomial of degree i, defined as

$$\frac{\mathrm{d}^i}{\mathrm{d}y^i}\phi(y) = (-1)^i J_i(y)\phi(y) . \tag{1.3.10}$$

The right-hand side of (1.3.8) resembles the frequentist Edgeworth expansion but has a much simpler derivation, as it only involves derivatives of the prior density and the log-likelihood function with respect to θ.

Let $P^{\overline{\pi}}\{\cdot|X\}$ be the posterior probability measure under the prior $\overline{\pi}(\cdot)$. Then, with $q(X,\theta)$ defined as in (1.3.2), from (1.3.8)–(1.3.10), we get

$$\begin{aligned}
E^{\overline{\pi}}\{q(X,\theta)|X\} &= P^{\overline{\pi}}\{n^{1/2}(\hat{\theta}-\theta)\le u|X\} \\
&= P^{\overline{\pi}}\{y \ge -uc^{1/2}|X\} \\
&= \int_{-uc^{1/2}}^{\infty} \phi(y)[1+n^{-1/2}\{G_1 J_1(y)+G_3 J_3(y)\}]\mathrm{d}y + o(n^{-1/2}) \\
&= \Phi(uc^{1/2})+n^{-1/2}\{G_1 + G_3(u^2 c-1)\}\phi(uc^{1/2}) + o(n^{-1/2}) \\
&= \Phi(uc^{1/2})+n^{-1/2}\{A_1 + A_3(u^2 c+2)\}\phi(uc^{1/2}) \\
&\quad + o(n^{-1/2}) ,
\end{aligned} \tag{1.3.11}$$

where $\Phi(\cdot)$ is the standard univariate normal cumulative distribution function.

Step 2: Let

$$I \equiv I(\theta) = E_\theta[\{\mathrm{d}\log f(X_1;\theta)/\mathrm{d}\theta\}^2] = -E_\theta\{\mathrm{d}^2 \log f(X_1;\theta)/\mathrm{d}\theta^2\}$$

be the per observation Fisher information at θ. Also define

$$L \equiv L(\theta) = E_\theta\{\mathrm{d}^3 \log f(X_1;\theta)/\mathrm{d}\theta^3\} ,$$

which, like I, is supposed to be a smooth function of θ. Now $E_\theta(\hat{\theta}) = \theta + o(n^{-1/2})$ and hence from (1.3.1), expanding $\ell''(\hat{\theta})$ about θ, one gets $E_\theta(c) = I + o(n^{-1/2})$. Similarly, $E_\theta(a) = L + o(n^{-1/2})$. Therefore, consideration of a Taylor's expansion about $(\hat{\theta},\ c,\ a) = (\theta,\ I,\ L)$ for the first two terms in the right-hand side of (1.3.11), upon substitution of (1.3.6) there, yields

$$\begin{aligned}
\lambda(\theta) &= E_\theta E^{\overline{\pi}}\{q(X,\theta)|X\} \\
&= \Phi(uI^{1/2})+n^{-1/2}\Big\{I^{-1/2}\frac{\overline{\pi}'(\theta)}{\overline{\pi}(\theta)} + \frac{1}{6}I^{-3/2}L(u^2 I+2)\Big\}\phi(uI^{1/2}) \\
&\quad + o(n^{-1/2}) ,
\end{aligned} \tag{1.3.12}$$

for θ in the interior of the support of $\overline{\pi}(\cdot)$. As a consequence of the Edgeworth assumptions and the uniformity property stated earlier about the P_θ−probability of the set $\overline{S}$, on which posterior expansions are being considered, the remainder in (1.3.12) is $o(n^{-1/2})$ uniformly over compact θ−sets in the interior of the support of $\overline{\pi}(\cdot)$.

Step 3: By (1.3.12),

$$\int \lambda(\theta)\overline{\pi}(\theta)d\theta = \int \left\{ \Phi(uI^{1/2}) + \frac{1}{6}n^{-1/2}I^{-3/2}L(u^2I + 2)\phi(uI^{1/2}) \right\} \overline{\pi}(\theta)d\theta$$

$$+ n^{-1/2} \int I^{-1/2}\overline{\pi}'(\theta)\phi(uI^{1/2})d\theta + o(n^{-1/2}) . \qquad (1.3.13)$$

The fact noted at the end of Step 2, in conjunction with the conditions on $\overline{\pi}(\cdot)$ that control $\log \overline{\pi}(\cdot)$ and its derivatives near the boundary of the support, facilitates passage from (1.3.12) to (1.3.13); cf. Ghosh, Sinha and Joshi (1982). Since $\overline{\pi}(\cdot)$ vanishes on the boundary of its support, upon integration by parts,

$$\int I^{-1/2}\overline{\pi}'(\theta)\phi(uI^{1/2})d\theta = - \int \left[\frac{d}{d\theta}\left\{ I^{-1/2}\phi(uI^{1/2}) \right\} \right] \overline{\pi}(\theta)d\theta .$$

Now suppose the support of $\overline{\pi}(\cdot)$ includes the true θ as an interior point. Then using the above in (1.3.13) and, thereafter, allowing $\overline{\pi}(\cdot)$ to converge weakly to the degenerate prior at the true θ, by (1.3.2) we get

$$P_\theta\{n^{1/2}(\widehat{\theta} - \theta) \le u\} = E_\theta\{q(X,\theta)\}$$

$$= \Phi(uI^{1/2}) + n^{-1/2}\left[\frac{1}{6}I^{-3/2}L(u^2I + 2)\phi(uI^{1/2}) \right.$$

$$\left. - \frac{d}{d\theta}\left\{ I^{-1/2}\phi(uI^{1/2}) \right\} \right] + o(n^{-1/2}) . \qquad (1.3.14)$$

The above argument is applicable whatever the true θ may be and, as such, (1.3.14) holds for all θ in the parameter space. This happens in the subsequent chapters as well whenever a frequentist expression is obtained as a result of Steps 1–3.

Using the regularity condition $dI/d\theta = -(L + M)$ (Bartlett, 1953), where

$$M = E_\theta[\{d\log f(X_1;\theta)/d\theta\}\{d^2\log f(X_1;\theta)/d\theta^2\}] ,$$

it can be seen that (1.3.14) is in agreement with what one obtains through a direct Edgeworth expansion, given for example in McCullagh (1987, Ch. 6). However, the use of a direct Edgeworth expansion warrants a stochastic expansion for $n^{1/2}(\widehat{\theta} - \theta)$ and then evaluation of approximate cumulants. The present shrinkage argument avoids all this algebra and simplifies the computation greatly.

Incidentally, posterior expansions such as (1.3.11) can be useful in other contexts too; see Woodroofe (1986) for an application to the study of the integrated averages of coverage probabilities of confidence intervals under sequential sampling.

2

Matching Priors for Posterior Quantiles

2.1 Introduction

The early literature on matching priors centered around those which ensure approximate frequentist validity of the posterior quantiles of a one-dimensional interest parameter (Welch and Peers, 1963; Peers, 1965). Even a major part of the recent research on matching priors has been on priors of this kind. In the present chapter, we review these developments. Specifically, we shall be considering priors $\pi(\cdot)$ for which the relation

$$P_\theta\{\theta_1 \leq \theta_1^{(1-\alpha)}(\pi, X)\} = 1 - \alpha + o(n^{-r/2}) , \qquad (2.1.1)$$

holds for $r = 1$ or 2 and for each α ($0 < \alpha < 1$). Here n is the sample size, $\theta = (\theta_1, \ldots, \theta_p)^T$ is an unknown parameter vector, θ_1 is the one-dimensional parameter of interest, $P_\theta\{\cdot\}$ is the frequentist probability measure under θ, and $\theta_1^{(1-\alpha)}(\pi, X)$ is the $(1-\alpha)$th posterior quantile of θ_1, under $\pi(\cdot)$, given the data X. Of course, we require (2.1.1) and its counterparts considered later in this chapter to hold for all possible θ as well, a point which is implicit throughout. Priors satisfying (2.1.1) for $r = 1$ or 2 are called *first* or *second order* matching priors, respectively. Clearly, they ensure that one-sided Bayesian credible sets of the form $(-\infty, \theta_1^{(1-\alpha)}(\pi, X)]$ for θ_1 have correct frequentist coverage as well, up to the order of approximation indicated in (2.1.1). As will be seen later, for $p \geq 2$, i.e., in the presence of nuisance parameters, a first order matching prior is not unique. The study of second order matching priors, which ensure correct frequentist coverage to a higher order of approximation, can help in significantly narrowing down the class of competing first order matching priors.

In the next section, we describe the setup in more detail and introduce the notation and preliminaries. Then in Section 2.3, an approximation for the posterior quantile $\theta_1^{(1-\alpha)}(\pi, X)$ is worked out. Section 2.4 deals with the characterization of first and second order matching priors via appropriate partial differential equations. Some important special cases are discussed in

Section 2.5. These include the case $p = 1$ (i.e., absence of nuisance parameters) and the situation where the nuisance parameters are chosen orthogonally to the interest parameter θ_1. We also focus attention to some general classes of models in this section. Further illustrative examples are presented in Section 2.6 and the issue of invariance, with respect to reparameterization, is taken up in Section 2.7. Results concerning general parametric functions, together with an application to Bayesian tolerance limits, are considered in Section 2.8. A related topic of matching alternative coverage probabilities has been discussed in Section 2.9. Incidentally, throughout Sections 2.2–2.6, we consider matching priors only in the sense of (2.1.1) even when this is not mentioned explicitly. Finally, in Section 2.10 we reconsider the examples presented in this chapter and comment on the propriety of the posteriors under the matching priors obtained in these examples.

2.2 Setup, notation and preliminaries

Consider a sequence $\{X_i\}, i \geq 1$, of i.i.d. possibly vector-valued random variables with common density $f(x; \theta)$, where the parameter vector $\theta = (\theta_1, \ldots, \theta_p)^T$ belongs to $\mathcal{R}^p$ or some open subset thereof and θ_1 is the parameter of interest. Let $X = (X_1, \ldots, X_n)^T$, where n is the sample size, and let $\widehat{\theta} = (\widehat{\theta}_1, \cdots, \widehat{\theta}_p)^T$ be the maximum likelihood estimator (MLE) of θ based on X.

We work under model assumptions similar to those in Section 1.3, with necessary changes so as to allow asymptotics up to the order of approximation $o(n^{-1})$ rather than $o(n^{-1/2})$ as considered there. Specifically, the model assumptions for the posterior and frequentist expansions considered in this and the subsequent chapters are as in Johnson (1970, pp. 852–853, with $K = 2$ in his notation) and Bhattacharya and Ghosh (1978, p. 439, with $s = 4$ in their notation) respectively. The latter assumptions entail the existence of a valid Edgeworth expansion, with margin of error $o(n^{-1})$, for the distribution of $n^{1/2}(\widehat{\theta} - \theta)$. These two sets of assumptions hold under wide generality for models belonging to the exponential or curved exponential families and also for many other models like Cauchy, Student's t, and so on. As a common feature of both sets, the per observation Fisher information matrix, namely I as defined below, is supposed to be positive definite for all θ. Indeed, between the two sets of assumptions, there is an overlapping part that governs the local behavior of the likelihood function near the true θ. This is basically similar to, but somewhat stronger than, the conditions that ensure the existence of a consistent and asymptotically normal solution to the likelihood equation. In addition, the Edgeworth assumptions in Bhattacharya and Ghosh (1978) involve conditions allowing expansions for the log-likelihood and the condition that each random variable X_i has a density. On the other hand, the additional assumptions in Johnson (1970) involve a condition on the behavior of $f(x; \theta)$ for θ bounded away from zero.

Let θ have a prior density $\pi(\cdot)$ which is positive and three times continuously differentiable over the entire parameter space. In case $\pi(\cdot)$ is not proper, we shall require that there is an n_0 (≥ 1) such that the posterior density of θ given $X = (X_1, \ldots, X_n)^T$ is proper, with P_θ−probability unity for all θ, when $n \geq n_0$. Note that $\pi(\cdot)$ is the prior of our main concern; later on, in Section 2.4, an auxiliary prior $\bar{\pi}(\cdot)$, analogous to the one in Section 1.3, will also be considered for frequentist computations relating to $\pi(\cdot)$. Any formal expansion for the posterior under $\pi(\cdot)$, as used in this and the subsequent chapters, is valid for sample points in a set S over which $\widehat{\theta}$ is well-defined and the observed information matrix C, as defined below, is positive definite. The P_θ−probability of S is $1 + o(n^{-1})$ uniformly on compact θ−sets in the parameter space. An explicit description of S can be given along the line of Bickel and Ghosh (1990, Section 3).

Let $\ell(\theta) = n^{-1} \sum_{i=1}^n \log f(X_i; \theta)$ and, with $D_j = \partial/\partial\theta_j$, for $1 \leq j, r, s, u \leq p$, let

$$\left.\begin{array}{ll} a_{jr} = \{D_j D_r \ell(\theta)\}_{\theta=\widehat{\theta}}\,, & c_{jr} = -a_{jr}\,, \\ a_{jrs} = \{D_j D_r D_s \ell(\theta)\}_{\theta=\widehat{\theta}}\,, & a_{jrsu} = \{D_j D_r D_s D_u \ell(\theta)\}_{\theta=\widehat{\theta}}\,, \end{array}\right\} \quad (2.2.1)$$

$$\left.\begin{array}{ll} V_j = D_j \log f(X_1; \theta)\,, & V_{jr} = D_j D_r \log f(X_1; \theta)\,, \\ & V_{jrs} = D_j D_r D_s \log f(X_1; \theta)\,, \end{array}\right\} \quad (2.2.2)$$

$$\left.\begin{array}{ll} I_{jr} = E_\theta(V_j V_r)\,, & L_{j,r,s} = E_\theta(V_j V_r V_s)\,, \\ L_{j,rs} = E_\theta(V_j V_{rs})\,, & L_{jrs} = E_\theta(V_{jrs})\,. \end{array}\right\} \quad (2.2.3)$$

As indicated above, the $p \times p$ matrix $C = ((c_{jr}))$ is positive definite over S. This is the per observation "observed" information matrix at $\widehat{\theta}$. Let $C^{-1} = ((c^{jr}))$ and, for $1 \leq j, r \leq p$, define

$$m_j = c^{j1}/c^{11}\,, \quad k^{jr} = c^{jr} - (c^{j1}c^{r1}/c^{11})\,. \quad (2.2.4)$$

Similarly, let $I = ((I_{jr}))$ be the per observation Fisher information matrix (i.e., "expected" information matrix) at θ. Define $I^{-1} = ((I^{jr}))$,

$$\tau^{jr} = I^{j1}I^{r1}/I^{11}\,, \quad \sigma^{jr} = I^{jr} - \tau^{jr}\,. \quad (2.2.5)$$

The quantities I_{jr}, I^{jr}, τ^{jr}, σ^{jr}, $L_{j,r,s}$ etc. are supposed to be smooth functions of θ. We also define, for $1 \leq j, r \leq p$,

$$\left.\begin{array}{ll} \pi_j(\theta) = D_j \pi(\theta)\,, & \pi_{jr}(\theta) = D_j D_r \pi(\theta)\,, \\ \widehat{\pi} = \pi(\widehat{\theta})\,, \quad \widehat{\pi}_j = \pi_j(\widehat{\theta})\,, & \widehat{\pi}_{jr} = \pi_{jr}(\widehat{\theta})\,. \end{array}\right\} \quad (2.2.6)$$

Let

$$h = (h_1, \ldots, h_p)^T = n^{1/2}(\theta - \widehat{\theta})\,. \quad (2.2.7)$$

We begin by finding an expansion for the posterior density of h. To that effect, as in Section 1.3, first note that the posterior density of θ under the prior $\pi(\cdot)$ equals

$$\pi(\theta|X) = \frac{\pi(\theta)\exp\{n\ell(\theta)\}}{\int \pi(\theta)\exp\{n\ell(\theta)\}d\theta}$$

$$= \frac{\pi(\theta)\exp[n\{\ell(\theta) - \ell(\widehat{\theta})\}]}{\int \pi(\theta)\exp[n\{\ell(\theta) - \ell(\widehat{\theta})\}]d\theta}.$$

Hence by (2.2.7), analogously to (1.3.3), the posterior density of h under the prior $\pi(\cdot)$ is given by

$$\pi^*(h|X) = \frac{b(h, X)}{\int b(h, X)dh}, \tag{2.2.8}$$

where

$$b(h, X) = \pi(\widehat{\theta} + n^{-1/2}h)\exp[n\{\ell(\widehat{\theta} + n^{-1/2}h) - \ell(\widehat{\theta})\}]. \tag{2.2.9}$$

From the definition of $\widehat{\theta}$, we have $\{D_j\ell(\theta)\}_{\theta=\widehat{\theta}} = 0$ for each j. Hence by (2.2.1) and (2.2.6), using Taylor's expansion about $\widehat{\theta}$,

$$\pi(\widehat{\theta} + n^{-1/2}h) = \widehat{\pi}\left\{1 + n^{-1/2}R_1(h) + \frac{1}{2}n^{-1}R_2(h)\right\} + o(n^{-1}), \tag{2.2.10}$$

and

$$n\{\ell(\widehat{\theta} + n^{-1/2}h) - \ell(\widehat{\theta})\} = -\frac{1}{2}c_{jr}h_jh_r + \frac{1}{6}n^{-1/2}R_3(h) + \frac{1}{24}n^{-1}R_4(h)$$
$$+ o(n^{-1}), \tag{2.2.11}$$

where

$$\left.\begin{array}{ll} R_1(h) = \widehat{\pi}_j h_j/\widehat{\pi}, & R_2(h) = \widehat{\pi}_{jr}h_jh_r/\widehat{\pi}, \\ R_3(h) = a_{jrs}h_jh_rh_s, & R_4(h) = a_{jrsu}h_jh_rh_sh_u. \end{array}\right\} \tag{2.2.12}$$

In (2.2.11), (2.2.12) and elsewhere in this monograph, unless otherwise specified, we follow the summation convention with sums over all repeated superscripts or subscripts ranging from 1 to p. Thus

$$c_{jr}h_jh_r = \sum_{j=1}^{p}\sum_{r=1}^{p} c_{jr}h_jh_r,$$

and so on.

Recalling that $C = ((c_{jr}))$, by (2.2.9)–(2.2.11),

$$b(h, X) = \widehat{\pi}\exp\left(-\frac{1}{2}h^T Ch\right)\left[1 + n^{-1/2}\left\{R_1(h) + \frac{1}{6}R_3(h)\right\}\right.$$
$$\left. + n^{-1}\left\{\frac{1}{2}R_2(h) + \frac{1}{6}R_1(h)R_3(h) + \frac{1}{24}R_4(h) + \frac{1}{72}\big(R_3(h)\big)^2\right\}\right]$$
$$+ o(n^{-1}). \tag{2.2.13}$$

Hence

$$\int b(h, X)\mathrm{d}h$$

$$= (2\pi)^{p/2}\{\det(C)\}^{-1/2}\widehat{\pi}\Big\{1 + n^{-1}\Big(\frac{1}{2}W_1 + \frac{1}{6}W_2 + \frac{1}{24}W_3 + \frac{1}{72}W_4\Big)\Big\}$$
$$+ o(n^{-1}), \qquad\qquad (2.2.14)$$

where the π in $(2\pi)^{p/2}$ is the usual transcendental number not to be confused with the prior $\pi(\cdot)$, and

$$\left.\begin{array}{ll} W_1 = \widehat{\pi}_{jr}c^{jr}/\widehat{\pi}, & W_2 = a_{jrs}\widehat{\pi}_u c^{(1)}_{jrsu}/\widehat{\pi}, \\[2mm] W_3 = a_{jrsu}c^{(1)}_{jrsu}, & W_4 = a_{jrs}a_{uvw}c^{(2)}_{jrsuvw}, \end{array}\right\} \qquad (2.2.15)$$

$$c^{(1)}_{jrsu} = c^{jr}c^{su} + c^{js}c^{ru} + c^{ju}c^{rs}, \qquad\qquad (2.2.16)$$

$$\begin{aligned} c^{(2)}_{jrsuvw} = \ & c^{jr}c^{su}c^{vw} + c^{jr}c^{sv}c^{uw} + c^{jr}c^{sw}c^{uv} \\ & + c^{js}c^{ru}c^{vw} + c^{js}c^{rv}c^{uw} + c^{js}c^{rw}c^{uv} \\ & + c^{ju}c^{rs}c^{vw} + c^{ju}c^{rv}c^{sw} + c^{ju}c^{rw}c^{sv} \\ & + c^{jv}c^{rs}c^{uw} + c^{jv}c^{ru}c^{sw} + c^{jv}c^{rw}c^{su} \\ & + c^{jw}c^{rs}c^{uv} + c^{jw}c^{ru}c^{sv} + c^{jw}c^{rv}c^{su}. \end{aligned} \qquad (2.2.17)$$

Since the quantities c^{jr}, a_{jrs} and a_{jrsu} are invariant under permutation of their superscripts or subscripts, by (2.2.15)–(2.2.17), one can simplify the expressions for W_2, W_3 and W_4 to get

$$\left.\begin{array}{ll} W_1 = \widehat{\pi}_{jr}c^{jr}/\widehat{\pi}, & W_2 = 3a_{jrs}\widehat{\pi}_u c^{jr}c^{su}/\widehat{\pi}, \\[2mm] W_3 = 3a_{jrsu}c^{jr}c^{su}, & W_4 = a_{jrs}a_{uvw}(9c^{jr}c^{su}c^{vw} + 6c^{ju}c^{rv}c^{sw}). \end{array}\right\} \quad (2.2.18)$$

By (2.2.8), (2.2.13) and (2.2.14), we get an expansion for the posterior density of h under the prior $\pi(\cdot)$ as

$$\begin{aligned} \pi^*(h|X) = \ & \phi_p(h; C^{-1})\Big[1 + n^{-1/2}\Big\{R_1(h) + \frac{1}{6}R_3(h)\Big\} \\ & + n^{-1}\Big\{\frac{1}{2}\Big(R_2(h) - W_1\Big) + \frac{1}{6}\Big(R_1(h)R_3(h) - W_2\Big) \\ & + \frac{1}{24}\Big(R_4(h) - W_3\Big) + \frac{1}{72}\Big((R_3(h))^2 - W_4\Big)\Big\}\Big] \\ & + o(n^{-1}), \end{aligned} \qquad (2.2.19)$$

where $\phi_p(\cdot; C^{-1})$ is the p−variate normal density with the null mean vector and dispersion matrix C^{-1}. The expansion (2.2.19), given originally in Ghosh and Mukerjee (1991), will be useful not only in this chapter but also in the subsequent chapters.

2.3 Posterior quantile

We now proceed to find an expression for $\theta_1^{(1-\alpha)}(\pi, X)$, the $(1-\alpha)$th posterior quantile of θ_1 under the prior $\pi(\cdot)$. This is done in Theorem 2.3.1 below. Since by (2.2.7), $h_1 = n^{1/2}(\theta_1 - \hat{\theta}_1)$, it will be helpful to first find an expression for the posterior density of h_1. This will be done by integrating $h_2, \ldots, h_p$ out from (2.2.19).

Let $h^* = (h_2, \ldots, h_p)^T$, $m = (m_2, \ldots, m_p)^T$ and define the square matrix K of order $p - 1$ as $K = ((k^{jr})) \ (2 \leq j, r \leq p)$, where the m_j and k^{jr} are as in (2.2.4). Observe that

$$\phi_p(h; C^{-1}) = \phi_1(h_1; c^{11})\phi_{p-1}(h^*; h_1 m, K) , \qquad (2.3.1)$$

where $\phi_1(\cdot; c^{11})$ is the univariate normal density with mean zero and variance c^{11}, and $\phi_{p-1}(\cdot; h_1 m, K)$ is the $(p-1)$-variate normal density with mean vector $h_1 m$ and dispersion matrix K. Also, from (2.2.4), for each j, r,

$$k^{jr} = c^{jr} - m_j m_r c^{11} . \qquad (2.3.2)$$

By (2.2.12), (2.2.18), (2.3.1) and (2.3.2), using the invariance of the a_{jrs} and k^{jr} under permutation of their subscripts or superscripts,

$$\int \phi_p(h; C^{-1})R_1(h)dh^* = \phi_1(h_1; c^{11})(\hat{\pi}_j m_j/\hat{\pi})h_1 ,$$

$$\int \phi_p(h; C^{-1})\{R_2(h) - W_1\}dh^* = \phi_1(h_1; c^{11})(\hat{\pi}_{jr} m_j m_r/\hat{\pi})(h_1^2 - c^{11}) ,$$

$$\int \phi_p(h; C^{-1})R_3(h)dh^* = \phi_1(h_1; c^{11})\{a_{jrs}(3k^{jr}m_s h_1 + m_j m_r m_s h_1^3)\} ,$$

$$\int \phi_p(h; C^{-1})\{R_1(h)R_3(h) - W_2\}dh^*$$
$$= \phi_1(h_1; c^{11})\Big\{(a_{jrs}\hat{\pi}_u/\hat{\pi})\Big(3k^{jr}k^{su} + 3k^{jr}m_s m_u h_1^2$$
$$+ 3k^{su}m_j m_r h_1^2 + m_j m_r m_s m_u h_1^4 - 3c^{jr}c^{su}\Big)\Big\}$$
$$= \phi_1(h_1; c^{11})\Big[(a_{jrs}\hat{\pi}_u/\hat{\pi})\Big\{3(k^{jr}m_s m_u + k^{su}m_j m_r)(h_1^2 - c^{11})$$
$$+ m_j m_r m_s m_u\Big(h_1^4 - 3(c^{11})^2\Big)\Big\}\Big] .$$

Furthermore, as $R_3(h)$, $R_4(h)$, W_3 and W_4 do not involve the prior $\pi(\cdot)$ or its derivatives, from (2.2.12) and (2.2.18), we similarly have

$$\int \phi_p(h; C^{-1})\{R_4(h) - W_3\}dh^*$$
$$= \phi_1(h_1; c^{11})\Big[B_1(X)(h_1^2 - c^{11}) + B_2(X)\{h_1^4 - 3(c^{11})^2\}\Big] ,$$

$$\int \phi_p(h; C^{-1})\left\{\left(R_3(h)\right)^2 - W_4\right\}dh^*$$
$$= \phi_1(h_1; c^{11})[B_3(X)(h_1^2 - c^{11}) + B_4(X)\{h_1^4 - 3(c^{11})^2\}$$
$$+ B_5(X)\{h_1^6 - 15(c^{11})^3\}],$$

where the quantities $B_i(X)$ $(1 \le i \le 5)$ are at most of order $O(1)$ and do not involve $\pi(\cdot)$ or its derivatives.

In view of the above, integrating out $h_2, \cdots, h_p$ from (2.2.19), after simplification, we get an expression for the posterior density of h_1 under the prior $\pi(\cdot)$ as

$$\tilde{\pi}(h_1|X) = \phi_1(h_1; c^{11})\Big[1 + n^{-1/2}\{A_1(\pi, X)h_1 + A_3(X)h_1^3\}$$
$$+ n^{-1}\Big\{A_2(\pi, X)(h_1^2 - c^{11}) + A_4(\pi, X)\Big(h_1^4 - 3(c^{11})^2\Big)$$
$$+ A_6(X)\Big(h_1^6 - 15(c^{11})^3\Big)\Big\}\Big] + o(n^{-1}), \qquad (2.3.3)$$

where

$$A_1(\pi, X) = A_{11}(\pi, X) + A_{12}(X), \quad A_{11}(\pi, X) = \hat{\pi}_j m_j/\hat{\pi}, \qquad (2.3.4)$$
$$A_{12}(X) = \frac{1}{2}a_{jrs}k^{jr}m_s, \quad A_3(X) = \frac{1}{6}a_{jrs}m_j m_r m_s, \quad (2.3.5)$$

$$\left.\begin{array}{l} A_2(\pi, X) = A_{21}(\pi, X) + A_{22}(X), \\ A_4(\pi, X) = A_{41}(\pi, X) + A_{42}(X), \end{array}\right\} \qquad (2.3.6)$$

$$A_{21}(\pi, X) = \frac{1}{2}(m_j m_r/\hat{\pi})(\hat{\pi}_{jr} + a_{jrs}k^{su}\hat{\pi}_u) + A_{11}(\pi, X)A_{12}(X), \quad (2.3.7)$$

$$A_{41}(\pi, X) = A_{11}(\pi, X)A_3(X), \qquad (2.3.8)$$

and like $A_{12}(X)$ and $A_3(X)$, the quantities $A_{22}(X)$, $A_{42}(X)$ and $A_6(X)$ are at most of order $O(1)$ and do not involve $\pi(\cdot)$ or its derivatives. The detailed expressions for these quantities will not be required in the sequel.

Some more notation will be needed before we present the main result of this section, namely, Theorem 2.3.1. Let z be the $(1 - \alpha)$th quantile of the standard univariate normal distribution and, as in (1.3.10) in Chapter 1, $J_i(\cdot)$ be the Hermite polynomial of degree i. Define

$$G_1(\pi) \equiv G_1(\pi, X) = A_1(\pi, X)(c^{11})^{1/2} + 3A_3(X)(c^{11})^{3/2}, \qquad (2.3.9)$$

$$G_3 \equiv G_3(X) = A_3(X)(c^{11})^{3/2}, \qquad (2.3.10)$$

$$G_2(\pi) \equiv G_2(\pi, X)$$
$$= A_2(\pi, X)c^{11} + 6A_4(\pi, X)(c^{11})^2 + 45A_6(X)(c^{11})^3, \quad (2.3.11)$$

$$G_4(\pi) \equiv G_4(\pi, X) = A_4(\pi, X)(c^{11})^2 + 15A_6(X)(c^{11})^3, \qquad (2.3.12)$$

$$G_6 \equiv G_6(X) = A_6(X)(c^{11})^3 , \qquad (2.3.13)$$

$$\beta_1(\alpha, \pi, X) = G_1(\pi) + G_3 J_2(z) , \qquad (2.3.14)$$

$$\beta_2(\alpha, \pi, X) = 2z\beta_1(\alpha, \pi, X)G_3 - \frac{1}{2}\{\beta_1(\alpha, \pi, X)\}^2 z$$
$$+ G_2(\pi)J_1(z) + G_4(\pi)J_3(z) + G_6 J_5(z) . \qquad (2.3.15)$$

Also, let $P^\pi\{\cdot|X\}$ be the posterior probability measure under the prior $\pi(\cdot)$.

Theorem 2.3.1. *Let*

$$\theta_1^{(1-\alpha)}(\pi, X) = \widehat{\theta}_1 + (n/c^{11})^{-1/2}\{z + n^{-1/2}\beta_1(\alpha, \pi, X) + n^{-1}\beta_2(\alpha, \pi, X)\} . \qquad (2.3.16)$$

Then

$$P^\pi\{\theta_1 \le \theta_1^{(1-\alpha)}(\pi, X)|X\} = 1 - \alpha + o(n^{-1}) .$$

Proof. Let

$$y = (c^{11})^{-1/2}h_1 = (n/c^{11})^{1/2}(\theta_1 - \widehat{\theta}_1) . \qquad (2.3.17)$$

From (1.3.10) recall that

$$\frac{d^i}{dy^i}\phi(y) = (-1)^i J_i(y)\phi(y) ,$$

where $\phi(\cdot)$ is the standard univariate normal density. Hence

$$J_1(y) = y , \quad J_2(y) = y^2 - 1 , \quad J_3(y) = y^3 - 3y , \quad J_4(y) = y^4 - 6y^2 + 3 ,$$

$$J_5(y) = y^5 - 10y^3 + 15y , \quad J_6(y) = y^6 - 15y^4 + 45y^2 - 15 ,$$

so that

$$y^3 = J_3(y) + 3J_1(y) , \quad y^4 - 3 = J_4(y) + 6J_2(y) ,$$

$$y^6 - 15 = J_6(y) + 15J_4(y) + 45J_2(y) .$$

Hence by (2.3.3) and (2.3.9)–(2.3.13), it is not hard to see that an expansion for the posterior density of y under the prior $\pi(\cdot)$ is given by

$$\pi_y(y|X) = \phi(y)\left[1 + n^{-1/2}\{G_1(\pi)J_1(y) + G_3 J_3(y)\}\right.$$

$$+ n^{-1}\{G_2(\pi)J_2(y) + G_4(\pi)J_4(y) + G_6 J_6(y)\}\bigg]$$

$$+ o(n^{-1}) . \qquad (2.3.18)$$

For notational simplicity, let $\beta_i = \beta_i(\alpha, \pi, X)(i = 1, 2)$. From (2.3.16)–(2.3.18) and the fact that $\int_{-\infty}^w J_1(y)\phi(y)dy = -\phi(w)$ and $\int_{-\infty}^w J_i(y)\phi(y)dy = -J_{i-1}(w)\phi(w)$, $i = 2, \ldots, 6$, one now obtains an expansion

$$P^\pi\{\theta_1 \leq \theta_1^{(1-\alpha)}(\pi, X)|X\}$$
$$= P^\pi\{y \leq z + n^{-1/2}\beta_1 + n^{-1}\beta_2|X\}$$
$$= \Phi(z + n^{-1/2}\beta_1 + n^{-1}\beta_2)$$
$$\quad - n^{-1/2}\phi(z + n^{-1/2}\beta_1)\{G_1(\pi) + G_3 J_2(z + n^{-1/2}\beta_1)\}$$
$$\quad - n^{-1}\phi(z)\{G_2(\pi)J_1(z) + G_4(\pi)J_3(z) + G_6 J_5(z)\} + o(n^{-1})$$
$$= \Phi(z) + n^{-1/2}\phi(z)\{\beta_1 - G_1(\pi) - G_3 J_2(z)\}$$
$$\quad + n^{-1}\phi(z)\Big[\beta_2 - 2z\beta_1 G_3 - \frac{1}{2}\beta_1^2 z + \beta_1 z\{G_1(\pi) + G_3 J_2(z)\}$$
$$\quad - G_2(\pi)J_1(z) - G_4(\pi)J_3(z) - G_6 J_5(z)\Big] + o(n^{-1}) . \qquad (2.3.19)$$

If one recalls the definition of z and employs (2.3.14) and (2.3.15) in (2.3.19), then the result is immediate. ♣

In view of Theorem 2.3.1, equation (2.3.16), with $\beta_1(\alpha, \pi, X)$ as in (2.3.14) and $\beta_2(\alpha, \pi, X)$ as in (2.3.15), gives an expression for the $(1 - \alpha)$th posterior quantile of θ_1, under $\pi(\cdot)$, up to the order of approximation $o(n^{-1})$.

2.4 Characterization of matching priors

We now calculate an expansion for the frequentist coverage probability $P_\theta\{\theta_1 \leq \theta_1^{(1-\alpha)}(\pi, X)\}$ with the objective of characterizing first and second order matching priors when θ_1 is the interest parameter; cf. (2.1.1). Steps 1–3 described in Section 1.2 greatly facilitate the computation. In order to apply these steps, we consider a proper auxiliary prior $\overline{\pi}(\cdot)$, satisfying the conditions in Bickel and Ghosh (1990), such that the support of $\overline{\pi}(\cdot)$ is a compact rectangle in the parameter space and the first partial derivatives of $\overline{\pi}(\cdot)$, like $\overline{\pi}(\cdot)$ itself, vanish on the boundary of the support. As before, $\overline{\pi}(\cdot)$ is positive in the interior of its support. Any formal expansion for the posterior under $\overline{\pi}(\cdot)$, as considered below and in the subsequent chapters, is over a set $\overline{S}$ with P_θ-probability $1 + o(n^{-1})$ uniformly on compact θ-sets in the interior of the support of $\overline{\pi}(\cdot)$. This is in the spirit of Section 1.3 and has similar implication.

Step 1: We first consider $P^{\overline{\pi}}\{\theta_1 \leq \theta_1^{(1-\alpha)}(\pi, X)|X\}$. Let y be defined as in (2.3.17). An expansion for the posterior density of y, under $\overline{\pi}(\cdot)$, is given by (2.3.18) with $G_1(\pi)$, $G_2(\pi)$ and $G_4(\pi)$ there replaced by $G_1(\overline{\pi})$, $G_2(\overline{\pi})$ and $G_4(\overline{\pi})$ respectively. Hence by (2.3.16), analogously to (2.3.19),

$$P^{\overline{\pi}}\{\theta_1 \leq \theta_1^{(1-\alpha)}(\pi, X)|X\}$$
$$= \Phi(z) + n^{-1/2}\phi(z)\{\beta_1(\alpha, \pi, X) - G_1(\overline{\pi}) - G_3 J_2(z)\}$$
$$\quad + n^{-1}\phi(z)\Big[\beta_2(\alpha, \pi, X) - 2z\beta_1(\alpha, \pi, X)G_3 - \frac{1}{2}\{\beta_1(\alpha, \pi, X)\}^2 z$$

$$+ \beta_1(\alpha, \pi, X)z\{G_1(\overline{\pi}) + G_3 J_2(z)\} - G_2(\overline{\pi})J_1(z)$$
$$- G_4(\overline{\pi})J_3(z) - G_6 J_5(z)\Big] + o(n^{-1}) .$$

Recalling the definition of z and employing (2.3.14) and (2.3.15) in the above, we get

$$P^{\overline{\pi}}\{\theta_1 \le \theta_1^{(1-\alpha)}(\pi, X)|X\}$$
$$= 1 - \alpha + n^{-1/2}\phi(z)\{G_1(\pi) - G_1(\overline{\pi})\}$$
$$+ n^{-1}\phi(z)\Big[\{G_2(\pi) - G_2(\overline{\pi})\}J_1(z) + \{G_4(\pi) - G_4(\overline{\pi})\}J_3(z)$$
$$- z\{G_1(\pi) - G_1(\overline{\pi})\}\{G_1(\pi) + G_3 J_2(z)\}\Big]$$
$$+ o(n^{-1}) . \tag{2.4.1}$$

But by (2.3.4), (2.3.6), (2.3.8)–(2.3.10) and (2.3.12),

$$G_4(\pi) - G_4(\overline{\pi}) = \{G_1(\pi) - G_1(\overline{\pi})\}G_3 .$$

Therefore, (2.4.1) yields

$$P^{\overline{\pi}}\{\theta_1 \le \theta_1^{(1-\alpha)}(\pi, X)|X\}$$
$$= 1 - \alpha + n^{-1/2}\phi(z)\{G_1(\pi) - G_1(\overline{\pi})\}$$
$$+ n^{-1}z\phi(z)[G_2(\pi) - G_2(\overline{\pi}) - \{G_1(\pi) - G_1(\overline{\pi})\}\{G_1(\pi) + 2G_3\}]$$
$$+ o(n^{-1}) . \tag{2.4.2}$$

Step 2: In this step one considers $E_\theta[P^{\overline{\pi}}\{\theta_1 \le \theta_1^{(1-\alpha)}(\pi, X)|X\}]$. Note that by (2.3.4) and (2.3.9),

$$G_1(\pi) - G_1(\overline{\pi}) = \{A_{11}(\pi, X) - A_{11}(\overline{\pi}, X)\}(c^{11})^{1/2}$$
$$= \Big(\frac{\widehat{\pi}_j}{\widehat{\pi}} - \frac{\widehat{\overline{\pi}}_j}{\widehat{\overline{\pi}}}\Big)m_j(c^{11})^{1/2} . \tag{2.4.3}$$

Also, by (2.3.4)–(2.3.11),

$$G_2(\pi) - G_2(\overline{\pi}) - \{G_1(\pi) - G_1(\overline{\pi})\}\{G_1(\pi) + 2G_3\}$$
$$= \{A_{21}(\pi, X) - A_{21}(\overline{\pi}, X)\}c^{11}$$
$$+ 6\{A_{11}(\pi, X) - A_{11}(\overline{\pi}, X)\}A_3(X)(c^{11})^2$$
$$- \{A_{11}(\pi, X) - A_{11}(\overline{\pi}, X)\}\{A_1(\pi, X) + 5A_3(X)c^{11}\}c^{11}$$
$$= \{A_{21}(\pi, X) - A_{21}(\overline{\pi}, X)\}c^{11}$$
$$+ \{A_{11}(\pi, X) - A_{11}(\overline{\pi}, X)\}\{A_3(X)c^{11} - A_1(\pi, X)\}c^{11}$$
$$= \frac{1}{2}m_j m_r\Big\{\Big(\frac{\widehat{\pi}_{jr}}{\widehat{\pi}} - \frac{\widehat{\overline{\pi}}_{jr}}{\widehat{\overline{\pi}}}\Big) + a_{jrs}k^{su}\Big(\frac{\widehat{\pi}_u}{\widehat{\pi}} - \frac{\widehat{\overline{\pi}}_u}{\widehat{\overline{\pi}}}\Big)\Big\}c^{11}$$
$$+ \Big\{\Big(\frac{\widehat{\pi}_u}{\widehat{\pi}} - \frac{\widehat{\overline{\pi}}_u}{\widehat{\overline{\pi}}}\Big)m_u\Big\}\Big(\frac{1}{6}a_{jrs}m_j m_r m_s c^{11} - \frac{\widehat{\pi}_j m_j}{\widehat{\pi}}\Big)c^{11} . \tag{2.4.4}$$

In view of (2.4.2)–(2.4.4), by (2.2.1)–(2.2.6), arguments similar to those in Section 1.3 show that

$$E_\theta[P^{\overline{\pi}}\{\theta_1 \leq \theta_1^{(1-\alpha)}(\pi, X)|X\}]$$

$$= 1 - \alpha + n^{-1/2}\phi(z)\Big\{\frac{\pi_j(\theta)}{\pi(\theta)} - \frac{\overline{\pi}_j(\theta)}{\overline{\pi}(\theta)}\Big\}I^{j1}(I^{11})^{-1/2}$$

$$+ n^{-1}z\phi(z)\Big[\frac{1}{2}\tau^{jr}\Big\{\frac{\pi_{jr}(\theta)}{\pi(\theta)} - \frac{\overline{\pi}_{jr}(\theta)}{\overline{\pi}(\theta)} + L_{jrs}\sigma^{su}\Big(\frac{\pi_u(\theta)}{\pi(\theta)} - \frac{\overline{\pi}_u(\theta)}{\overline{\pi}(\theta)}\Big)\Big\}$$

$$+ \Big\{\Big(\frac{\pi_u(\theta)}{\pi(\theta)} - \frac{\overline{\pi}_u(\theta)}{\overline{\pi}(\theta)}\Big)\frac{I^{u1}}{I^{11}}\Big\}\frac{H(\pi,\theta)}{\pi(\theta)}\Big] + o(n^{-1}), \qquad (2.4.5)$$

for θ in the interior of the support of $\overline{\pi}(\cdot)$, where

$$H(\pi,\theta) = \frac{1}{6}L_{jrs}\tau^{jr}I^{s1}\pi(\theta) - I^{j1}\pi_j(\theta). \qquad (2.4.6)$$

Step 3: This is the final step in the calculation of the frequentist coverage probability $P_\theta\{\theta_1 \leq \theta_1^{(1-\alpha)}(\pi, X)\}$. We now suppose that the support of $\overline{\pi}(\cdot)$ contains the true θ as an interior point. Recall also that the first partial derivatives of $\overline{\pi}(\cdot)$, like $\overline{\pi}(\cdot)$ itself, vanish on the boundary of the support which is a compact rectangle. We integrate $E_\theta[P^{\overline{\pi}}\{\theta_1 \leq \theta_1^{(1-\alpha)}(\pi, X)|X\}]$ by parts with respect to $\overline{\pi}(\cdot)$ and then allow $\overline{\pi}(\cdot)$ to converge weakly to the degenerate prior at the true θ. By (2.4.5), as in Section 1.3, this yields

$$P_\theta\{\theta_1 \leq \theta_1^{(1-\alpha)}(\pi, X)\} = 1 - \alpha + n^{-1/2}\frac{\phi(z)}{\pi(\theta)}\Delta_1(\pi, \theta)$$

$$+ n^{-1}\frac{z\phi(z)}{\pi(\theta)}\Delta_2(\pi, \theta) + o(n^{-1}), \qquad (2.4.7)$$

for all θ, where

$$\Delta_1(\pi, \theta) = \pi_j(\theta)I^{j1}(I^{11})^{-1/2} + \pi(\theta)D_j\{I^{j1}(I^{11})^{-1/2}\}$$

$$= D_j\{\pi(\theta)I^{j1}(I^{11})^{-1/2}\}, \qquad (2.4.8)$$

and

$$\Delta_2(\pi, \theta) = \frac{1}{2}\tau^{jr}\pi_{jr}(\theta) - \frac{1}{2}\pi(\theta)D_jD_r(\tau^{jr}) + \frac{1}{2}\tau^{jr}L_{jrs}\sigma^{su}\pi_u(\theta)$$

$$+ \frac{1}{2}\pi(\theta)D_u(\tau^{jr}L_{jrs}\sigma^{su}) + \pi_u(\theta)(I^{u1}/I^{11})\{H(\pi,\theta)/\pi(\theta)\}$$

$$+ \pi(\theta)D_u[(I^{u1}/I^{11})\{H(\pi,\theta)/\pi(\theta)\}]$$

$$= \frac{1}{2}\tau^{jr}\pi_{jr}(\theta) - \frac{1}{2}\pi(\theta)D_jD_r(\tau^{jr})$$

$$+ D_u\Big\{\frac{1}{2}\pi(\theta)\tau^{jr}L_{jrs}\sigma^{su} + (I^{u1}/I^{11})H(\pi,\theta)\Big\}. \qquad (2.4.9)$$

But by (2.2.5) and (2.4.6),

$$(I^{u1}/I^{11})H(\pi,\theta) = \frac{1}{6}\pi(\theta)\tau^{jr}L_{jrs}\tau^{su} - \tau^{ju}\pi_j(\theta) \,,$$

while

$$D_u\{\tau^{ju}\pi_j(\theta)\} = \pi_j(\theta)D_u(\tau^{ju}) + \tau^{ju}\pi_{ju}(\theta)$$
$$= \frac{1}{2}\pi_j(\theta)D_r(\tau^{jr}) + \frac{1}{2}\pi_r(\theta)D_j(\tau^{jr}) + \tau^{jr}\pi_{jr}(\theta) \,.$$

Hence the expression for $\Delta_2(\pi,\theta)$, as given in (2.4.9), can be simplified as

$$\begin{aligned}
\Delta_2(\pi,\theta) &= \frac{1}{2}\tau^{jr}\pi_{jr}(\theta) - \frac{1}{2}\pi(\theta)D_jD_r(\tau^{jr}) + \frac{1}{2}D_u\{\pi(\theta)\tau^{jr}L_{jrs}\sigma^{su}\} \\
&\quad + \frac{1}{6}D_u\{\pi(\theta)\tau^{jr}L_{jrs}\tau^{su}\} - \frac{1}{2}\pi_j(\theta)D_r(\tau^{jr}) \\
&\quad - \frac{1}{2}\pi_r(\theta)D_j(\tau^{jr}) - \tau^{jr}\pi_{jr}(\theta) \\
&= \frac{1}{6}D_u\{\pi(\theta)\tau^{jr}L_{jrs}(3\sigma^{su} + \tau^{su})\} - \frac{1}{2}\Big\{\pi(\theta)D_jD_r(\tau^{jr}) \\
&\quad + \pi_j(\theta)D_r(\tau^{jr}) + \pi_r(\theta)D_j(\tau^{jr}) + \tau^{jr}\pi_{jr}(\theta)\Big\} \\
&= \frac{1}{6}D_u\{\pi(\theta)\tau^{jr}L_{jrs}(3\sigma^{su} + \tau^{su})\} - \frac{1}{2}D_jD_r\{\pi(\theta)\tau^{jr}\} \,. \quad (2.4.10)
\end{aligned}$$

From (2.4.7), it is clear that

$$P_\theta\{\theta_1 \le \theta_1^{(1-\alpha)}(\pi,X)\} = 1 - \alpha + o(n^{-1/2})$$

for all α, if and only if the prior $\pi(\cdot)$ satisfies $\Delta_1(\pi,\theta) = 0$. Furthermore,

$$P_\theta\{\theta_1 \le \theta_1^{(1-\alpha)}(\pi,X)\} = 1 - \alpha + o(n^{-1})$$

for all α, if and only if $\pi(\cdot)$ satisfies, in addition, $\Delta_2(\pi,\theta) = 0$. Hence, in consideration of (2.4.8) and (2.4.10) and with reference to ensuring approximate frequentist validity of the posterior quantiles of θ_1, we get the following theorem which is the main result of this chapter.

Theorem 2.4.1. *(a) A prior $\pi(\cdot)$ is first order probability matching if and only if it satisfies the partial differential equation*

$$D_j\{\pi(\theta)I^{j1}(I^{11})^{-1/2}\} = 0 \,. \tag{2.4.11}$$

(b) The prior $\pi(\cdot)$ is second order probability matching if and only if it satisfies, in addition, the partial differential equation

$$\frac{1}{3}D_u\{\pi(\theta)\tau^{jr}L_{jrs}(3\sigma^{su} + \tau^{su})\} - D_jD_r\{\pi(\theta)\tau^{jr}\} = 0 \,. \tag{2.4.12}$$

♣

Part (a) of Theorem 2.4.1 was proved originally by Peers (1965) and part (b) by Mukerjee and Ghosh (1997). Earlier, several authors studied important special cases of this theorem. Applications to general classes of models were considered by others. Some of this work will be discussed in the next section.

While concluding this section, we refer to Ghosal (1999) for results on probability matching priors in some nonregular cases where the support of the density underlying the model depends on the parameter of interest. We also refer to Rousseau (2000) who obtained results on matching priors in the discrete case via continuity corrections.

2.5 Special cases

We now focus attention to the consequences of Theorem 2.4.1 in some important special cases and for a few general classes of models. For ease in presentation, we split the section into several subsections.

2.5.1 The case $p = 1$

We first consider matching priors in one-parameter models. For $p = 1$, i.e., in the absence of any nuisance parameter, we have $\theta = \theta_1$. In this case, both θ and I are scalars. Then the first order matching equation (2.4.11) reduces to

$$\frac{\mathrm{d}}{\mathrm{d}\theta}\left\{\pi(\theta)/I^{1/2}\right\} = 0 ,$$

leading to the unique solution

$$\pi(\theta) \propto I^{1/2} , \tag{2.5.1}$$

which is the well-known Jeffreys' (1961) prior and, incidentally, also the reference prior (Bernardo, 1979). Thus, in this case, Jeffreys' prior is the unique first order matching prior. Furthermore, for $p = 1$, by (2.2.3) and (2.2.5), it is not hard to check that (2.4.12) reduces to

$$\frac{1}{3}\frac{\mathrm{d}}{\mathrm{d}\theta}\left\{\pi(\theta)L_{111}/I^2\right\} - \frac{\mathrm{d}^2}{\mathrm{d}\theta^2}\left\{\pi(\theta)/I\right\} = 0 ,$$

i.e., $\dfrac{1}{3}\left\{\pi(\theta)L_{111}/I^2\right\} - \dfrac{\mathrm{d}}{\mathrm{d}\theta}\left\{\pi(\theta)/I\right\} = \text{constant} . \tag{2.5.2}$

Now, for Jeffreys' prior, given by (2.5.1), the left-hand side of (2.5.2) simplifies to

$$\frac{1}{3}I^{-3/2}L_{111} - \frac{\mathrm{d}}{\mathrm{d}\theta}I^{-1/2} = \frac{1}{6}I^{-3/2}L_{1,1,1} , \tag{2.5.3}$$

using the standard regularity conditions (Bartlett, 1953)

$$\frac{\mathrm{d}}{\mathrm{d}\theta}I = -(L_{1,11} + L_{111}) , \tag{2.5.4}$$

and

$$L_{111} + 3L_{1,11} + L_{1,1,1} = 0 \,, \tag{2.5.5}$$

both of which are consequences of differentiation under the integral sign.

Summarizing (2.5.1)–(2.5.3), in the present context we get Theorem 2.5.1 below. This is an early result on matching priors. It was obtained originally by Welch and Peers (1963) and explored further by Peers (1965) and Stein (1985).

Theorem 2.5.1. *(a) For $p = 1$, Jeffreys' prior, given by $\pi(\theta) \propto I^{1/2}$, is the unique first order probability matching prior.*

(b) Furthermore, it is also second order probability matching if and only if $I^{-3/2}L_{1,1,1}$ is a constant free from θ. ♣

From (2.2.3) recall that

$$I = E_\theta[\{d\log f(X_1;\theta)/d\theta\}^2] \,, \quad L_{1,1,1} = E_\theta[\{d\log f(X_1;\theta)/d\theta\}^3] \,.$$

Since $d\log f(X_1;\theta)/d\theta$ has zero mean under θ, the quantity $I^{-3/2}L_{1,1,1}$ can be interpreted as the skewness coefficient of $d\log f(X_1;\theta)/d\theta$ under θ. Thus, by Theorem 2.5.1(b), Jeffreys' prior is second order probability matching if and only if this skewness coefficient does not depend on θ.

As an illustration, consider the one-parameter location model

$$f(x;\theta) = f^*(x - \theta), \quad x \in \mathcal{R}^1 \,, \tag{2.5.6}$$

where $\theta \in \mathcal{R}^1$ and $f^*(\cdot)$ is a density with support $\mathcal{R}^1$. Then both I and $L_{1,1,1}$ are constants which do not involve θ. As such, by Theorem 2.5.1, Jeffreys' prior, given by $\pi(\theta) = $ constant, is second order probability matching. In fact, the matching property of Jeffreys' prior is exact for the model (2.5.6). This result, again due to Welch and Peers (1963), is presented in Theorem 2.5.2 below. Earlier, Lindley (1958) reported the same result under the existence of a single sufficient statistic for θ.

Theorem 2.5.2. *Under the one-parameter location model (2.5.6), Jeffreys' prior, given by $\pi(\theta) = $ constant, is exact probability matching.*

Proof. Let $\theta^{(1-\alpha)}(\pi, X) = \theta^*$, say, be the $(1 - \alpha)$th posterior quantile of θ, given $X = (X_1, \cdots, X_n)^T$, under Jeffreys' prior $\pi(\theta) = $ constant. Then by (2.5.6), writing $u = X_1 - \theta$, we get

$$1 - \alpha = \frac{\int_{-\infty}^{\theta^*}\{\prod_{i=1}^n f^*(X_i - \theta)\}d\theta}{\int_{-\infty}^{\infty}\{\prod_{i=1}^n f^*(X_i - \theta)\}d\theta}$$

$$= \int_{X_1-\theta^*}^{\infty} \psi(u; X_2 - X_1, \cdots, X_n - X_1)du \,, \tag{2.5.7}$$

where for any real numbers $\gamma_2, \cdots, \gamma_n$,

$$\psi(u; \gamma_2, \cdots, \gamma_n) = \frac{f^*(u) \prod_{i=2}^n f^*(u + \gamma_i)}{\int_{-\infty}^\infty f^*(u)\{\prod_{i=2}^n f^*(u + \gamma_i)\}du} . \tag{2.5.8}$$

For any given $\gamma_2, \cdots, \gamma_n$, from (2.5.8) note that $\psi(\cdot; \gamma_2, \cdots, \gamma_n)$ is a density with support $\mathcal{R}^1$. Let $q_\alpha(\gamma_2, \cdots, \gamma_n)$ be the αth quantile of this density. Then by (2.5.7),

$$X_1 - \theta^* = q_\alpha(X_2 - X_1, \cdots, X_n - X_1) . \tag{2.5.9}$$

We now consider the frequentist probability $P_\theta(\theta \leq \theta^*)$. By (2.5.9),

$$\begin{aligned} P_\theta(\theta \leq \theta^*) &= P_\theta(X_1 - \theta \geq X_1 - \theta^*) \\ &= P_\theta\{U \geq q_\alpha(U_2, \cdots, U_n)\} , \end{aligned} \tag{2.5.10}$$

where $U = X_1 - \theta$ and $U_i = X_i - X_1$ ($2 \leq i \leq n$). Note that the joint density of $U, U_2, \cdots, U_n$ does not involve θ and is given by, say,

$$\widetilde{f}(u, u_2, \cdots, u_n) = f^*(u) \prod_{i=2}^n f^*(u + u_i) ,$$

with support $\mathcal{R}^n$. Hence the conditional probability density function of U given $U_2, \cdots, U_n$ is given by $\psi(u; u_2, \cdots, u_n)$; cf. (2.5.8). Recalling the definition of $q_\alpha(U_2, \cdots, U_n)$, it follows that given $U_2, \cdots, U_n$, the conditional frequentist probability of the event that

$$U \geq q_\alpha(U_2, \cdots, U_n)$$

is $1 - \alpha$ which does not involve $U_2, \cdots, U_n$. Therefore, the same must hold for the corresponding unconditional probability, so that by (2.5.10), $P_\theta(\theta \leq \theta^*) = 1 - \alpha$, completing the proof.   ♣

Consider now the one-parameter scale model

$$f(x; \theta) = \frac{1}{\theta} f^*\left(\frac{x}{\theta}\right) , \tag{2.5.11}$$

where $\theta > 0$ and $f^*(\cdot)$ is a density with support either $\mathcal{R}^1$ or $[0, \infty)$. Then $I \propto \theta^{-2}$ and $L_{1,1,1} \propto \theta^{-3}$, so that by Theorem 2.5.1, Jeffreys' prior, given by $\pi(\theta) \propto \theta^{-1}$, is second order probability matching. With the scale model (2.5.11) too, in fact, it can be shown that the matching property of Jeffreys' prior is exact. The proof proceeds along the line of Theorem 2.5.2 with the modification that one has to consider ratios instead of differences.

Even beyond the standard one-parameter location or scale models, Jeffreys' prior can enjoy the second order matching property. On the other hand, there can be models where the condition in Theorem 2.5.1(b) does not hold and consequently no second order matching prior is available. Illustrative examples follow.

Example 2.5.1. Consider the multivariate normal regression model

$$f(x;\theta) = \Big\{ \prod_{j=1}^{t} \phi(x^{(j)}) \Big\} \phi\Big(x^{(t+1)} - \sum_{j=1}^{t} \rho_j(\theta) x^{(j)} \Big) \,,$$

where $x = (x^{(1)}, \cdots, x^{(t+1)})^T \in \mathcal{R}^{t+1}$, $\phi(\cdot)$ is as usual the standard univariate normal density, and $\rho_1(\theta), \cdots, \rho_t(\theta)$ are smooth functions of a scalar parameter θ belonging to $\mathcal{R}^1$ or some open subset thereof. Then

$$I = \sum_{j=1}^{t} \Big\{ \frac{d\rho_j(\theta)}{d\theta} \Big\}^2$$

and $L_{1,1,1} = 0$, identically in θ, so that by Theorem 2.5.1, Jeffreys' prior enjoys the second order matching property.   ♣

Example 2.5.2. Consider the bivariate normal model with zero means, unit variances and correlation coefficient θ, where $|\theta| < 1$. Then

$$I = \frac{1 + \theta^2}{(1 - \theta^2)^2} \,, \quad L_{1,1,1} = -\frac{2\theta(3 + \theta^2)}{(1 - \theta^2)^3} \,.$$

Here $I^{-3/2} L_{1,1,1}$ is not free from θ. Hence by Theorem 2.5.1, Jeffreys' prior is only first order probability matching and no second order matching prior is available.   ♣

2.5.2 The case $p = 2$

We now discuss two-parameter models. Then $p = 2$ which corresponds to the situation where both the interest and nuisance parameters are one-dimensional and covers many models of interest such as the location-scale family considered below. By (2.2.5), here

$$\left.\begin{array}{l} I^{11} = Q \,, \quad I^{12} = I^{21} = -Q\zeta, \quad I^{22} = QI_{11}/I_{22} \,, \\ \tau^{11} = Q \,, \quad \tau^{12} = \tau^{21} = -Q\zeta, \quad \tau^{22} = Q\zeta^2 \,, \\ \sigma^{11} = 0 \,, \quad \sigma^{12} = \sigma^{21} = 0 \,, \quad \sigma^{22} = I_{22}^{-1} \,, \end{array}\right\} \qquad (2.5.12)$$

where

$$Q = (I_{11} - I_{22}^{-1} I_{12}^2)^{-1} \,, \quad \zeta = I_{12}/I_{22} \,. \qquad (2.5.13)$$

We also recall that the quantities L_{jrs} are invariant under permutation of their subscripts. Hence using (2.5.12), the partial differential equations (2.4.11) and (2.4.12), for first and second order probability matching, can be expressed as

$$D_1\{\pi(\theta)Q^{1/2}\} - D_2\{\pi(\theta)Q^{1/2}\zeta\} = 0 \,, \qquad (2.5.14)$$

and

$$\frac{1}{3}D_1\{\pi(\theta)Q^2(L_{111} - 3L_{112}\zeta + 3L_{122}\zeta^2 - L_{222}\zeta^3)\}$$

$$- \frac{1}{3}D_2\{\pi(\theta)Q^2\zeta(L_{111} - 3L_{112}\zeta + 3L_{122}\zeta^2 - L_{222}\zeta^3)\}$$

$$+ D_2\{\pi(\theta)QI_{22}^{-1}(L_{112} - 2L_{122}\zeta + L_{222}\zeta^2)\}$$

$$- D_1^2\{\pi(\theta)Q\} + 2D_1D_2\{\pi(\theta)Q\zeta\} - D_2^2\{\pi(\theta)Q\zeta^2\} = 0, \quad (2.5.15)$$

respectively.

The second order matching condition (2.5.15) for the case $p = 2$ is due to Mukerjee and Dey (1993). Their result, though not stated explicitly in this form, is equivalent to (2.5.15). Levine and Casella (2003) studied analytical as well as numerical methods for solving the first order matching condition (2.5.14). Substantial simplification occurs in (2.5.14) and (2.5.15) if $I_{12} = 0$, identically in θ, which corresponds to orthogonal parameterization and entails $\zeta = 0$. The consequences of such simplification were studied by Lee (1989) for $p = 2$ and will be taken up in the next subsection under the framework of general p.

For illustration, we consider the location-scale model

$$f(x;\theta) = \frac{1}{\theta_2}f^*\left(\frac{x - \theta_1}{\theta_2}\right), \; x \in \mathcal{R}^1, \quad (2.5.16)$$

where $\theta_1 \in \mathcal{R}^1, \theta_2 > 0$, and $f^*(\cdot)$ is a density with support $\mathcal{R}^1$. First suppose θ_1, the location parameter, is of interest while the scale parameter θ_2 is the nuisance parameter. Here for $j, r, s = 1, 2$,

$$I_{jr} \propto \theta_2^{-2}, \; L_{jrs} \propto \theta_2^{-3}. \quad (2.5.17)$$

Hence by (2.5.13), $Q \propto \theta_2^2$ and ζ is a constant. It can now be checked that the prior

$$\pi(\theta) \propto \theta_2^{-1} \quad (2.5.18)$$

satisfies both (2.5.14) and (2.5.15) and is, therefore, second order probability matching. Interchanging the roles of θ_1 and θ_2 in (2.5.14) and (2.5.15), one can also verify that the same prior enjoys the second order matching property when θ_2 is the interest parameter and θ_1 is the nuisance parameter. We note in this context that under the usual rectangular compactification of the parameter space, (2.5.18) represents the reference prior as well whether θ_1 or θ_2 is the interest parameter (Berger and Bernardo, 1989).

It is of interest to examine if the prior in (2.5.18) is the unique second order matching prior for θ_1 or θ_2. To that effect, continuing with the location-scale model (2.5.16), we now suppose that

$$I_{12} = 0, \; L_{122} = 0, \; L_{112} \neq 0, \quad (2.5.19)$$

a feature that arises commonly – e.g., with the univariate normal or Cauchy settings. Then using (2.5.13) and (2.5.17),

$$I_{11} = r_1/\theta_2^2 \,, \ I_{22} = r_2/\theta_2^2 \,, \ Q = \theta_2^2/r_1 \,, \ \zeta = 0 \,,$$

$$L_{111} = r_3/\theta_2^3 \,, \ L_{112} = r_4/\theta_2^3 \,,$$

where $r_1, \cdots, r_4$ are constants, $r_4 \neq 0$ and $r_1, r_2 > 0$. Hence under (2.5.19), the partial differential equations (2.5.14) and (2.5.15) reduce to

$$D_1\{\pi(\theta)\theta_2\} = 0 \qquad\qquad (2.5.20)$$

and

$$\frac{1}{3}r_3 r_1^{-2} D_1\{\pi(\theta)\theta_2\} + r_4(r_1 r_2)^{-1} D_2\{\pi(\theta)\theta_2\} - r_1^{-1} D_1^2\{\pi(\theta)\theta_2^2\} = 0 \,. \ (2.5.21)$$

Since $r_4 \neq 0$, it follows that the prior in (2.5.18) is the unique solution to (2.5.20) and (2.5.21). Thus under the commonly arising conditions (2.5.19), it is the unique second order matching prior when the location parameter θ_1 is of interest.

We now turn to the situation where interest lies in the scale parameter θ_2. Then, under (2.5.19), it can be similarly seen that priors of the form $\pi(\theta) = d(\theta_1)/\theta_2$, where $d(\cdot)(> 0)$ is any smooth function, constitute the class of second order matching priors. The prior in (2.5.18) is indeed of this form its lack of uniqueness is a consequence of the fact that $L_{122} \equiv 0$; vide (2.5.19).

Interestingly, for the univariate normal model with mean θ_1 and standard deviation θ_2, the prior given by (2.5.18) is exact probability matching irrespective of whether θ_1 or θ_2 is the interest parameter. This point, noted among others by Guttman (1970, Ch. 7), Box and Tiao (1973, Ch. 2) and Sun and Ye (1996), is a consequence of the fact that under both the frequentist and posterior (with respect to the prior in (2.5.18)) settings

(a) $n^{1/2}(\overline{X} - \theta_1)/s$ follows the t-distribution with $n - 1$ degrees of freedom, and

(b) $(n - 1)s^2/\theta_2^2$ follows the chi-square distribution with $n - 1$ degrees of freedom.

In the above,

$$\overline{X} = n^{-1}\sum_{i=1}^{n} X_i \ \text{and} \ s^2 = (n - 1)^{-1}\sum_{i=1}^{n}(X_i - \overline{X})^2 \,,$$

where $X_1, X_2, \ldots$ are i.i.d. each having the normal distribution under consideration.

2.5.3 Orthogonal parameterization

In the last subsection, we had hinted that for $p = 2$, the study of matching priors can get substantially simplified when $I_{12} = 0$, identically in θ. With reference to general p, we shall now discuss the implications of this phenomenon, known as orthogonal parameterization, in some detail. Let

$$I_{1j} = 0 \ (2 \le j \le p) , \tag{2.5.22}$$

identically in θ. Then $I^{1j} = 0$, $2 \le j \le p$, and by (2.2.5),

$$\tau^{11} = I^{11} = I_{11}^{-1} ;$$

$$\tau^{jr} = 0 , \text{ if } (j,r) \ne (1,1) ;$$

$$\sigma^{jr} = 0 , \text{ if either } j = 1 \text{ or } r = 1 ;$$

$$\sigma^{jr} = I^{jr} , \text{ if } j \ge 2 \text{ and } r \ge 2 .$$

Hence the partial differential equations (2.4.11) and (2.4.12), for first and second order probability matching, can be expressed as

$$D_1\{\pi(\theta)I_{11}^{-1/2}\} = 0 , \tag{2.5.23}$$

and

$$\sum_{s=2}^{p}\sum_{u=2}^{p}D_u\{\pi(\theta)I_{11}^{-1}I^{su}L_{11s}\} + \frac{1}{3}D_1\{\pi(\theta)I_{11}^{-2}L_{111}\} - D_1^2\{\pi(\theta)I_{11}^{-1}\} = 0 , \tag{2.5.24}$$

respectively.

A prior $\pi(\cdot)$ satisfies (2.5.23) and is hence first order probability matching if and only if it is of the form

$$\pi(\theta) = d(\theta^{(2)})I_{11}^{1/2} , \tag{2.5.25}$$

where $d(\cdot)(> 0)$ is any smooth function of $\theta^{(2)} = (\theta_2, \cdots, \theta_p)^T$. This result is due to Tibshirani (1989); Nicolaou (1993) also proved it using another approach.

With a prior of the form (2.5.25), the left-hand side of (2.5.24) equals

$$\sum_{s=2}^{p}\sum_{u=2}^{p}D_u\{d(\theta^{(2)})I_{11}^{-1/2}I^{su}L_{11s}\} + d(\theta^{(2)})D_1\left\{\frac{1}{3}I_{11}^{-3/2}L_{111} - D_1I_{11}^{-1/2}\right\} ,$$

and, analogously to (2.5.3),

$$\frac{1}{3}I_{11}^{-3/2}L_{111} - D_1I_{11}^{-1/2} = \frac{1}{6}I_{11}^{-3/2}L_{1,1,1} .$$

Hence by (2.5.24), a prior of the form (2.5.25) is second order probability matching if and only if

$$\sum_{s=2}^{p}\sum_{u=2}^{p}D_u\{d(\theta^{(2)})I_{11}^{-1/2}I^{su}L_{11s}\} + \frac{1}{6}d(\theta^{(2)})D_1\{I_{11}^{-3/2}L_{1,1,1}\} = 0 . \tag{2.5.26}$$

Since the interest parameter θ_1 is one-dimensional, in principle, orthogonal parameterization is always possible. In other words, as noted in Cox and Reid (1987) (see also Huzurbazar, 1950), one can always choose the $(p-1)$-dimensional nuisance parameter $(\theta_2, \cdots, \theta_p)^T$ such that (2.5.22) holds identically in θ. Also, we shall see later in Section 2.7 that the matching priors considered in this chapter are invariant of the parameterization. In particular, they are invariant with respect to the choice of $\theta_2, \cdots, \theta_p$ when interest lies in θ_1. At this stage, the reader may be curious as to why then we did not assume, at the outset, an orthogonal parameterization which could have resulted in a simpler direct derivation of the matching conditions (2.5.23) and (2.5.24), instead of proceeding via the conditions (2.4.11) and (2.4.12) of Theorem 2.4.1. The reason is that although in principle an orthogonal parameterization is always possible in the present setup, its explicit determination calls for solving certain partial differential equations (see Cox and Reid, 1987) which can sometimes be intractable; see Cox and Reid (1993). This is why, we thought it appropriate to present the main result of this chapter, namely Theorem 2.4.1, and the associated matching conditions (2.4.11) and (2.4.12) in a form that does not require orthogonal parameterization. At any rate, the invariance argument, hinted above and discussed later in Section 2.7, justifies the reparameterization used in the next and many of the subsequent examples.

Example 2.5.3. (Mukerjee and Ghosh, 1997) Consider the bivariate normal model with means μ_1, μ_2, variances γ_1^2, γ_2^2 and correlation coefficient ρ. The parameters are all unknown and interest lies in the regression coefficient $\rho\gamma_2/\gamma_1$. Reparameterize as

$$\theta_1 = \rho\gamma_2/\gamma_1 \,, \ \theta_2 = \gamma_2^2(1-\rho^2) \,, \ \theta_3 = \gamma_1^2 \,, \ \theta_4 = \mu_1 \,, \ \theta_5 = \mu_2 \,, \quad (2.5.27)$$

where $\theta_1, \theta_4, \theta_5 \in \mathcal{R}^1$ and $\theta_2, \theta_3 > 0$. Then θ_1 becomes the interest parameter. One can check that (2.5.27) is an orthogonal parameterization in the sense of (2.5.22) and that $I_{11} = \theta_3/\theta_2$. Hence by (2.5.25), first order matching is achieved if and only if

$$\pi(\theta) = d(\theta^{(2)})(\theta_3/\theta_2)^{1/2} \,, \quad (2.5.28)$$

where as before $\theta^{(2)} = (\theta_2, \cdots, \theta_5)^T$. Also, here

$$I_{22} = \frac{1}{2}\theta_2^{-2} \,, \ I_{23} = I_{24} = I_{25} = 0 \,,$$

$$L_{112} = \theta_3/\theta_2^2 \,, \ L_{113} = L_{114} = L_{115} = L_{1,1,1} = 0 \,.$$

Therefore, (2.5.26) holds if and only if $d(\theta^{(2)})$ is of the form $\tilde{d}(\theta^{(3)})/\theta_2^{1/2}$, where $\theta^{(3)} = (\theta_3, \theta_4, \theta_5)^T$. Coupled with (2.5.28), this implies that $\pi(\cdot)$ is second order matching if and only if

$$\pi(\theta) = d^*(\theta^{(3)})/\theta_2 \,, \quad (2.5.29)$$

where $d^*(\cdot)(>0)$ is any smooth function of $\theta^{(3)}$. Thus a considerable reduction of the class of first order matching priors is possible via second order matching.

With reference to the original $(\mu_1, \mu_2, \gamma_1, \gamma_2, \rho)$−parameterization, priors of the form

$$\pi^*(\mu_1, \mu_2, \gamma_1, \gamma_2, \rho) = \{\gamma_1^r \gamma_2^s (1 - \rho^2)^t\}^{-1} \qquad (2.5.30)$$

are of natural interest in the setup of this example and one may wish to know for what choice of the real numbers r, s and t these priors are first or second order probability matching. By (2.5.27), the Jacobian of the transformation $(\mu_1, \mu_2, \gamma_1, \gamma_2, \rho) \to (\theta_1, \cdots, \theta_5)$ is $\frac{1}{4}(\theta_2 + \theta_1^2 \theta_3)^{-1}$. Hence, under the θ−parameterization, (2.5.30) gets transformed to

$$\pi(\theta) = \frac{(\theta_2 + \theta_1^2 \theta_3)^{t - \frac{1}{2}s - 1}}{4\theta_3^{\frac{1}{2}r} \theta_2^t} \ .$$

Comparing the above with (2.5.28) and (2.5.29), it follows that a prior of the form (2.5.30) is first order probability matching if and only if $t = \frac{1}{2}s + 1$, and it possesses the second order matching property if and only if in addition $t = 1$. ♣

Example 2.5.4. (Mukerjee and Dey, 1993) This concerns the ratio of independent exponential means. Let

$$f(x; \theta) = \frac{1}{\mu_1(\theta)\mu_2(\theta)} \exp\left[-\left\{\frac{x^{(1)}}{\mu_1(\theta)} + \frac{x^{(2)}}{\mu_2(\theta)}\right\}\right], \ x^{(1)}, \ x^{(2)} > 0 \ ,$$

where $x = (x^{(1)}, x^{(2)})^T$, $\mu_1(\theta) = \theta_1^{1/2}\theta_2$, $\mu_2(\theta) = \theta_1^{-1/2}\theta_2$, and $\theta_1, \theta_2 > 0$. Note that the interest parameter θ_1 equals the ratio $\mu_1(\theta)/\mu_2(\theta)$ and that $\mu_1(\theta), \mu_2(\theta) > 0$. Here we have orthogonal parameterization. Furthermore,

$$I_{11} = \frac{1}{2}\theta_1^{-2} \ , \ I_{22} = 2\theta_2^{-2} \ , \ L_{1,1,1} = 0 \ , \ L_{112} = \frac{1}{2}(\theta_1^2 \theta_2)^{-1} \ .$$

Hence by (2.5.25), first order matching is achieved if and only if $\pi(\theta) = d(\theta_2)/\theta_1$, whereas by (2.5.26) such a prior is second order matching if and only if, in addition, $d(\theta_2) \propto \theta_2^{-1}$. Thus $\pi(\theta) \propto (\theta_1\theta_2)^{-1}$ is the unique second order matching prior in this example. This is also the reference prior under the usual rectangular compactification of the parameter space (Berger and Bernardo, 1989). ♣

The special case $L_{11s} = 0$ $(2 \le s \le p)$, which arises for instance in Example 2.5.4 when the roles of θ_1 and θ_2 are interchanged, deserves attention. Then (2.5.26) reduces to the model condition

$$D_1\left(I_{11}^{-3/2} L_{1,1,1}\right) = 0 \ .$$

Hence either each or none of the first order matching priors is second order matching, depending on whether this model condition holds or not; cf. Theorem 2.5.1(b). Interestingly, along the line of DiCiccio and Stern (1994) one can check that under orthogonal parameterization, $L_{11s} = 0$ $(2 \le s \le p)$ if and only if

$$E_\theta\{\widehat{\theta}^{(2)}(\theta_1)\} - E_\theta\{\widehat{\theta}^{(2)}\} = o(n^{-1}) \,,$$

for every θ, where $\widehat{\theta}^{(2)}(\theta_1)$ is the MLE of $\theta^{(2)}$ given θ_1 and $\widehat{\theta}^{(2)}$ is the overall MLE of $\theta^{(2)}$. A transparent discussion of this issue, via a Laplace expansion for the posterior density, is available in Reid (2003).

At this stage, we note that the Jeffreys' prior $\pi(\theta) \propto \{\det(I)\}^{1/2}$ may or may not enjoy the probability matching property in the presence of nuisance parameters. For instance, the unique second order matching prior in Example 2.5.4 is also Jeffreys' prior. On the other hand, for any location-scale model, such as univariate normal, that satisfies (2.5.19), Jeffreys' prior is not even first order matching when interest lies in the scale parameter. The latter point is clear from the discussion that follows (2.5.19).

2.5.4 Two general classes of models

We now discuss matching priors with reference to two general classes of models both admitting orthogonal parameterization. Following Sun and Ye (1996), we first consider a two-parameter exponential family given by a density of the form

$$f(x; \theta) = w(x) \exp\{\theta_1 u_1(x) - \theta_1 g_2'(\theta_2) u_2(x) - \psi(\theta_1, \theta_2)\} \,, \qquad (2.5.31)$$

where $w(\cdot)$, $u_1(\cdot)$ and $u_2(\cdot)$ do not involve θ_1, θ_2,

$$\psi(\theta_1, \theta_2) = -\theta_1\{\theta_2 g_2'(\theta_2) - g_2(\theta_2)\} + g_1(\theta_1) \,, \qquad (2.5.32)$$

$\theta_1 < 0$, $g_1''(\theta_1) > 0$, $g_2''(\theta_2) > 0$, and $g_1(\cdot)$ and $g_2(\cdot)$ are smooth functions. The above family was discussed originally by Bar-Lev and Reiser (1982) from frequentist considerations. Important special cases of this family are the univariate normal, gamma and inverse Gaussian distributions. Here

$$I_{11} = g_1''(\theta_1) \,, \quad I_{22} = -\theta_1 g_2''(\theta_2) \,, \quad I_{12} = 0 \,. \qquad (2.5.33)$$

Hence by (2.5.25), if θ_1 is the interest parameter then any prior of the form

$$\pi(\theta) = d(\theta_2)\{g_1''(\theta_1)\}^{1/2} \,, \qquad (2.5.34)$$

is first order probability matching. On the other hand, if instead θ_2 is the interest parameter then in a similar manner any prior of the form

$$\pi(\theta) = d(\theta_1)\{-\theta_1 g_2''(\theta_2)\}^{1/2} \qquad (2.5.35)$$

enjoys this property. In either (2.5.34) or (2.5.35), $d(\cdot)(> 0)$ is any smooth function. As noted in Sun and Ye (1996) (see also Datta and Ghosh, M., 1995a) here the reference prior is given by

$$\pi(\theta) = \{g_1''(\theta_1)g_2''(\theta_2)\}^{1/2} , \qquad (2.5.36)$$

whether θ_1 or θ_2 is of interest. Comparing with (2.5.34) and (2.5.35), clearly the reference prior enjoys the first order matching property for both θ_1 and θ_2.

Continuing with the model (2.5.31), we now consider second order matching priors. From (2.5.31) and (2.5.32) one gets

$$L_{111} = -g_1'''(\theta_1) , \quad L_{222} = 2\theta_1 g_2'''(\theta_2) ,$$

$$L_{112} = 0 , \quad L_{122} = g_2''(\theta_2) .$$

Regularity conditions similar to (2.5.4) and (2.5.5) yield

$$L_{1,1,1} = 2L_{111} + 3D_1 I_{11} , \quad L_{2,2,2} = 2L_{222} + 3D_2 I_{22} .$$

Hence we also get

$$L_{1,1,1} = g_1'''(\theta_1) , \quad L_{2,2,2} = \theta_1 g_2'''(\theta_2) .$$

First suppose θ_1 is the interest parameter. Then, using the above, a first order matching prior $\pi(\cdot)$ as in (2.5.34) also satisfies the second order matching condition (2.5.26) if and only if

$$D_1[\{g_1''(\theta_1)\}^{-3/2} g_1'''(\theta_1)] = 0 ,$$

i.e., if and only if

$$D_1^2[\{g_1''(\theta_1)\}^{-1/2}] = 0 . \qquad (2.5.37)$$

Note that (2.5.37) is a condition on the model rather than the prior. As such, if (2.5.37) holds then all first order matching priors are second order matching; otherwise, no second order matching prior is available. Since $L_{112} = 0$, this is as per the discussion below Example 2.5.4.

Interchanging the roles of θ_1 and θ_2 in (2.5.26), it can similarly be seen that, with θ_2 as the interest parameter, a first order matching prior given by (2.5.35) also enjoys the second order matching property if and only if

$$\frac{(-\theta_1)^{1/2}}{d(\theta_1)} D_1 \left\{ \frac{d(\theta_1)}{(-\theta_1)^{1/2} g_1''(\theta_1)} \right\} = -\frac{1}{3} \{g_2''(\theta_2)\}^{-1/2} D_2^2[\{g_2''(\theta_2)\}^{-1/2}]$$

i.e., if and only if there exists a constant M such that

$$\frac{1}{3} \{g_2''(\theta_2)\}^{-1/2} D_2^2[\{g_2''(\theta_2)\}^{-1/2}] = -M \qquad (2.5.38)$$

and

$$d(\theta_1) \propto (-\theta_1)^{1/2} g_1''(\theta_1) \exp\{M g_1'(\theta_1)\} . \qquad (2.5.39)$$

Observe that (2.5.38) is again a condition on the model rather than the prior. If (2.5.38) holds then (2.5.35) and (2.5.39) jointly determine a unique second order matching prior for θ_2; otherwise, no such second order matching prior is available.

The findings in (2.5.34)–(2.5.39) are due to Sun and Ye (1996). Some minor typos in their paper have been rectified in (2.5.38) and (2.5.39). The next two examples show applications of (2.5.34)–(2.5.39) to the inverse Gaussian and gamma models. Since the univariate normal model has already been discussed in subsection 2.5.2, it is no more considered here.

Example 2.5.5. (Sun and Ye, 1996) Consider the inverse Gaussian model given by the density

$$\{\lambda_1/(2\pi x^3)\}^{1/2} \exp\left\{ -\frac{1}{2}\lambda_1 x(x^{-1} - \lambda_2)^2 \right\}, \ x > 0 \,,$$

where $\lambda_1, \lambda_2 > 0$ and π is the usual transcendental number (not to be confused with a prior). The above is of the form represented by (2.5.31) and (2.5.32) with

$$\theta_1 = -\frac{1}{2}\lambda_1 \,, \ \theta_2 = \lambda_2^{-1}, \ g_1(\theta_1) = -\frac{1}{2}\log(-2\theta_1) \,, \ g_2(\theta_2) = \theta_2^{-1} \,,$$

$$u_1(x) = x^{-1} \,, \ u_2(x) = x \,.$$

Note that θ_1 and θ_2 are one-to-one functions of λ_1 and λ_2 respectively. Here

$$g_1''(\theta_1) = \frac{1}{2}\theta_1^{-2} \,, \ g_2''(\theta_2) = 2\theta_2^{-3} \,. \tag{2.5.40}$$

Hence among the two model conditions (2.5.37) and (2.5.38) only the first one holds. Thus if θ_1 is the interest parameter then all priors of the form (2.5.34) enjoy not only the first order but also the second order matching property. On the other hand, if interest lies in θ_2 then priors of the form (2.5.35) are first order probability matching and no second order matching prior is available. By (2.5.36) and (2.5.40), the reference prior $\pi(\theta) \propto (-\theta_1)^{-1}\theta_2^{-3/2}$ is first order matching whether θ_1 or θ_2 is of interest.   ♣

Example 2.5.6. (Sun and Ye, 1996) Consider the gamma model given by the density

$$\lambda_1^{\lambda_1} x^{\lambda_1-1} \exp(-\lambda_1 x/\lambda_2)/\{\lambda_2^{\lambda_1} \Gamma(\lambda_1)\} \,, \ x > 0 \,,$$

where $\lambda_1, \lambda_2 > 0$ and $\Gamma(\cdot)$ is the gamma function. The above is of the form given by (2.5.31) and (2.5.32) with

$$\theta_1 = -\lambda_1 \,, \ \theta_2 = \lambda_2 \,, \ g_1(\theta_1) = -\theta_1 + \theta_1 \log(-\theta_1) + \log\Gamma(-\theta_1) \,,$$

$$g_2(\theta_2) = -\log\theta_2 \,, \ u_1(x) = -\log x \,, \ u_2(x) = x \,.$$

Here among the model conditions (2.5.37) and (2.5.38) the first one does not hold while the second one holds with $M = 0$. Therefore, no second order matching prior exists if θ_1 is of interest. On the other hand, following (2.5.35) and (2.5.39), one gets a unique second order matching prior when θ_2 is the interest parameter.   ♣

Following Garvan and Ghosh (1997), we now consider another general two-parameter family given by a density of the form

$$f(x;\theta) = g(x,\theta_2)\exp\{\theta_2 u(x,\theta_1)\}\,, \tag{2.5.41}$$

where θ_2 is either always positive or always negative, and both $g(\cdot,\cdot)$ and $u(\cdot,\cdot)$ are smooth functions. The above represents the dispersion model of Jorgensen (1992, Ch. 1; 1997, Ch. 1) and, with roles of θ_1 and θ_2 interchanged, includes the family shown earlier in (2.5.31) and (2.5.32).

Orthogonal parameterization holds under (2.5.41). Hence from (2.5.25), or its dual with θ_1 and θ_2 interchanged, one gets, as before, first order matching priors for θ_1 or θ_2 respectively. Turning to the issue of second order matching, under (2.5.41),

$$L_{112} = -\,\theta_2^{-1}I_{11}\,,\ L_{122} = 0\,. \tag{2.5.42}$$

Because of the first identity in (2.5.42), the second order matching condition (2.5.26) for θ_1 slightly simplifies to

$$-D_2\{d(\theta_2)\theta_2^{-1}I_{11}^{1/2}I_{22}^{-1}\} + \frac{1}{6}d(\theta_2)D_1(I_{11}^{-3/2}L_{1,1,1}) = 0\,. \tag{2.5.43}$$

If instead θ_2 is the interest parameter then interchanging the roles of θ_1 and θ_2 in (2.5.26) and using the second identity in (2.5.42), the second order matching condition reduces to

$$D_2(I_{22}^{-3/2}L_{2,2,2}) = 0\,. \tag{2.5.44}$$

Note that (2.5.44), like (2.5.37) above, is a model condition and has similar implication. An illustrative example follows.

Example 2.5.7. (Garvan and Ghosh, 1997) Consider the Student's t-model given by the density

$$f(x;\theta) = \frac{1}{B(\frac{1}{2},\theta_2 - \frac{1}{2})\{1 + (x - \theta_1)^2\}^{\theta_2}}\,,\ x \in \mathcal{R}^1\,,$$

where $\theta_1 \in \mathcal{R}^1$, $\theta_2 > \frac{1}{2}$, and $B(\cdot,\cdot)$ is the beta function. The above is clearly of the form (2.5.41). Here

$$I_{11} = \frac{\theta_2(2\theta_2 - 1)}{\theta_2 + 1}\,,\ I_{22} = D_2^2\log B\left(\frac{1}{2},\theta_2 - \frac{1}{2}\right)\,,\ L_{1,1,1} = 0\,. \tag{2.5.45}$$

Hence by (2.5.25) and its dual with θ_1 and θ_2 interchanged,

$$\pi(\theta) \propto \left\{D_2^2\log B\left(\frac{1}{2},\theta_2 - \frac{1}{2}\right)\right\}^{1/2}$$

is the unique prior which enjoys the first order matching property for both θ_1 and θ_2. Following Garvan and Ghosh (1997), this is also the reference prior whether θ_1 or θ_2 is of interest. Also, by (2.5.25), (2.5.43) and (2.5.45),

$$\pi(\theta) \propto \theta_2 D_2^2 \log B\left(\frac{1}{2}, \theta_2 - \frac{1}{2}\right)$$

is the unique second order matching prior when θ_1 is the interest parameter. On the other hand, the model condition (2.5.44) does not hold in this example and, therefore, no second order matching prior for θ_2 is available. ♣

2.6 Further examples

Example 2.6.1. (Mukerjee and Dey, 1993) This concerns the ratio of two independent normal means and corresponds to a simpler version of the Fieller (1954) and Creasy (1954) problem. Let the model be represented by the density

$$\phi(x^{(1)} - \mu_1)\phi(x^{(2)} - \mu_2), \quad x^{(1)}, \, x^{(2)} \in \mathcal{R}^1,$$

where μ_1, $\mu_2(> 0)$ are unknown parameters. As before, $\phi(\cdot)$ represents the standard univariate normal density. Interest lies in the ratio μ_1/μ_2. Reparameterize as

$$\theta_1 = \mu_1/\mu_2, \quad \theta_2 = (\mu_1^2 + \mu_2^2)^{1/2}, \tag{2.6.1}$$

where $\theta_1, \theta_2 > 0$. Then θ_1 is the parameter of interest and one can check that (2.6.1) is an orthogonal parameterization. Furthermore,

$$I_{11} = \theta_2^2/(\theta_1^2 + 1)^2, \quad I_{22} = 1, \quad L_{1,1,1} = 0, \quad L_{112} = -\theta_2/(\theta_1^2 + 1)^2.$$

Hence by (2.5.25), first order matching is achieved if and only if $\pi(\theta) = d(\theta_2)\theta_2/(\theta_1^2 + 1)$, whereas by (2.5.26) such a prior is second order matching if and only if, in addition, $d(\theta_2)$ is a constant. Thus $\pi(\theta) \propto \theta_2/(\theta_1^2 + 1)$ is the unique second order matching prior. Interestingly, under the original (μ_1, μ_2)–parameterization, this gets transformed to the flat prior. ♣

Example 2.6.2. (Tibshirani, 1989) Continuing with the setup of the last example, now suppose interest lies in the product $\mu_1\mu_2$. Reparameterize as

$$\theta_1 = \mu_1\mu_2, \quad \theta_2 = \mu_2^2 - \mu_1^2, \tag{2.6.2}$$

i.e.,

$$\mu_1 = \left[\frac{1}{2}\{(4\theta_1^2 + \theta_2^2)^{1/2} - \theta_2\}\right]^{1/2}, \quad \mu_2 = \left[\frac{1}{2}\{(4\theta_1^2 + \theta_2^2)^{1/2} + \theta_2\}\right]^{1/2},$$

where $\theta_1 > 0, \theta_2 \in \mathcal{R}^1$. Then θ_1 is the parameter of interest and it is not hard to see that (2.6.2) is an orthogonal parameterization. Furthermore,

$$I_{11} = 4I_{22} = (4\theta_1^2 + \theta_2^2)^{-1/2}, \quad L_{1,1,1} = 0, \quad L_{112} = \frac{1}{2}\theta_2(4\theta_1^2 + \theta_2^2)^{-3/2}. \tag{2.6.3}$$

By (2.5.25), a prior $\pi(\cdot)$ is first order matching if and only if it is of the form

$$\pi(\theta) = d(\theta_2)(4\theta_1^2 + \theta_2^2)^{-1/4} . \qquad (2.6.4)$$

Such a prior is also second order matching if and only if $d(\theta_2)$ satisfies (2.5.26) which, in view of (2.6.3), reduces to

$$D_2\{d(\theta_2)\theta_2(4\theta_1^2 + \theta_2^2)^{-3/4}\} = 0 .$$

Clearly, the above equation does not admit any solution for $d(\theta_2)$. Thus no second order matching prior is available in this example.

Taking $d(\theta_2)$ as constant in (2.6.4), one gets the first order matching prior $\pi(\theta) \propto (4\theta_1^2 + \theta_2^2)^{-1/4}$. Under the original (μ_1, μ_2)−parameterization, this is proportional to $(\mu_1^2 + \mu_2^2)^{1/2}$ which is a reference prior (Berger and Bernardo, 1989). ♣

Example 2.6.3. This example concerns the Behrens–Fisher problem and follows the line of Ghosh and Kim (2001). Let the model be represented by the density

$$\frac{1}{\gamma_1\gamma_2}\phi\Big(\frac{x^{(1)} - \mu_1}{\gamma_1}\Big)\phi\Big(\frac{x^{(2)} - \mu_2}{\gamma_2}\Big) , \quad x^{(1)} , \; x^{(2)} \in \mathcal{R}^1 ,$$

where μ_1, μ_2 $(\in \mathcal{R}^1)$ and γ_1, γ_2 (> 0) are unknown parameters. Interest lies in the difference $\mu_1 - \mu_2$. Reparameterize as

$$\theta_1 = \mu_1 - \mu_2 , \quad \theta_2 = \frac{(\mu_1/\gamma_1^2) + (\mu_2/\gamma_2^2)}{(1/\gamma_1^2) + (1/\gamma_2^2)} , \quad \theta_3 = \gamma_1 , \quad \theta_4 = \gamma_2 , \quad (2.6.5)$$

where $\theta_1, \theta_2 \in \mathcal{R}^1$, and $\theta_3, \theta_4 > 0$. Then θ_1 is the parameter of interest and one can check that (2.6.5) is an orthogonal parameterization. Also

$$I_{11} = (\theta_3^2 + \theta_4^2)^{-1} , \quad I^{33} = \frac{1}{2}\theta_3^2 , \quad I^{34} = 0 , \quad I^{44} = \frac{1}{2}\theta_4^2 ,$$

$$L_{1,1,1} = L_{112} = 0 , \quad L_{113} = 2\theta_3/(\theta_3^2 + \theta_4^2)^2 , \quad L_{114} = 2\theta_4/(\theta_3^2 + \theta_4^2)^2 ,$$

and none of I^{22}, I^{23} and I^{24} involves θ_2. Hence by (2.5.25), first order matching is achieved if and only if $\pi(\theta) = d(\theta^{(2)})(\theta_3^2 + \theta_4^2)^{-1/2}$, where $\theta^{(2)} = (\theta_2, \theta_3, \theta_4)^T$. Moreover, it can be seen that $d(\theta^{(2)}) \propto (\theta_3^2 + \theta_4^2)^{3/2}/(\theta_3\theta_4)^3$ satisfies second order matching condition (2.5.26). Hence $\pi(\theta) \propto (\theta_3^2 + \theta_4^2)/(\theta_3\theta_4)^3$ is a second order matching prior. ♣

Example 2.6.4. (Mukerjee and Ghosh, 1997) We continue with the setup of the last example but suppose that γ_1 and γ_2 are equal, their common value γ (> 0) being unknown. Interest lies in $(\mu_1 - \mu_2)/\gamma$. Reparameterize as

$$\theta_1 = (\mu_1 - \mu_2)/\gamma , \quad \theta_2 = \frac{1}{2}(\mu_1 + \mu_2) , \quad \theta_3 = \{8\gamma^2 + (\mu_1 - \mu_2)^2\}^{1/2} , \quad (2.6.6)$$

where $\theta_1, \theta_2 \in \mathcal{R}^1$ and $\theta_3 > 0$. Then θ_1 is the parameter of interest and it can be seen that (2.6.6) is an orthogonal parameterization. Here

$$I_{11} = 4/(8 + \theta_1^2) , \quad I_{23} = 0 , \quad I_{33} = (8 + \theta_1^2)/(2\theta_3^2) ,$$

$$L_{1,1,1} = -16\theta_1(12 + \theta_1^2)/(8 + \theta_1^2)^3 , \quad L_{112} = 0 , \quad L_{113} = 32/\{\theta_3(8 + \theta_1^2)^2\} .$$

Therefore, by (2.5.25), first order matching is achieved if and only if

$$\pi(\theta) = d(\theta^{(2)})(8 + \theta_1^2)^{-1/2} , \tag{2.6.7}$$

where $\theta^{(2)} = (\theta_2, \theta_3)^T$. Furthermore, by (2.5.26), such a prior is second order matching if and only if $d(\theta^{(2)})$ is free from θ_3, i.e., if and only if the prior is of the form

$$\pi(\theta) = d^*(\theta_2)(8 + \theta_1^2)^{-1/2} , \tag{2.6.8}$$

where $d^*(\theta_2)(> 0)$ is any smooth function of θ_2.

With reference to the original (μ_1, μ_2, γ)−parameterization, priors having the structure

$$\pi^*(\mu_1, \mu_2, \gamma) = \left[\left\{8 + \left(\frac{\mu_1 - \mu_2}{\gamma}\right)^2\right\}^{s_1} \gamma^{s_2}\right]^{-1} , \tag{2.6.9}$$

where s_1 and s_2 are real numbers, have received attention in the literature (Datta and Ghosh, J.K., 1995a; Ghosh and Yang, 1996). Under the θ−parameterization (2.6.6), priors of this kind get transformed to

$$\pi(\theta) = \{(8 + \theta_1^2)^{s_1+1-\frac{1}{2}s_2}\theta_3^{s_2-1}\}^{-1}. \tag{2.6.10}$$

Comparing the above with (2.6.7) and (2.6.8), it follows that a prior of the form (2.6.9) is first order matching if and only if $s_2 = 2s_1+1$ (Ghosh and Yang, 1996), and it enjoys the second order matching property if and only if $s_1 = 0$, $s_2 = 1$. Interestingly, with $s_1 = 0$, $s_2 = 1$ in (2.6.9), one gets a prior that is proportional to γ^{-1}; cf. (2.5.18). By (2.6.10), under the θ−parameterization, this prior corresponds to the choice $d^*(\theta_2) = $ constant in (2.6.8).   ♣

Example 2.6.5. (Mukerjee and Dey, 1993) Consider a version of the exponential regression model of Cox and Reid (1987) as given by the density

$$f(x; \theta) = \prod_{j=1}^{t} \left[\theta_2^{-1} \exp(-\theta_1 z_j) \exp\{-\theta_2^{-1}x^{(j)}e^{-\theta_1 z_j}\}\right] , \quad x^{(1)} , \ldots , x^{(t)} > 0 ,$$

where $x = (x^{(1)}, \ldots, x^{(t)})^T, \theta_1 \in \mathcal{R}^1, \theta_2 > 0, t(\geq 2)$ is an integer, and $z_1, \ldots, z_t$ are constants, not all zeros, satisfying $z_1 + \ldots + z_t = 0$. As usual, θ_1 is the parameter of interest. Here we have orthogonal parameterization. Furthermore,

$$I_{11} = \sum_{j=1}^{t} z_j^2 , \quad I_{22} = t/\theta_2^2 , \quad L_{1,1,1} = 2\sum_{j=1}^{t} z_j^3 , \quad L_{112} = \left(\sum_{j=1}^{t} z_j^2\right)/\theta_2 .$$

Since $z_1, \ldots, z_t$ are constants, by (2.5.25), a prior $\pi(\theta)$ is first order matching if and only if it is of the form $\pi(\theta) = d(\theta_2)$. Also, by (2.5.26), such a prior is second order matching if and only if $d(\theta_2) \propto \theta_2^{-1}$. Thus $\pi(\theta) \propto \theta_2^{-1}$ is the unique second order matching prior. Datta and Ghosh (1996) noted that this is the reference prior as well. Also, by using group invariance argument, these authors provided an alternative derivation of this prior.   ♣

Example 2.6.6. (Sun, 1997) Consider the Weibull model given by the density

$$(\mu_1/\mu_2)(x/\mu_2)^{\mu_1-1} \exp\{-(x/\mu_2)^{\mu_1}\} , \quad x > 0 ,$$

where $\mu_1, \mu_2 \ (> 0)$ are unknown parameters. Interest lies in the shape parameter μ_1. Reparameterize as

$$\theta_1 = \mu_1 , \quad \theta_2 = \mu_2 e^{w/\mu_1} , \tag{2.6.11}$$

where $w = \int_0^\infty (u \log u) e^{-u} du$ and $\theta_1, \theta_2 > 0$. Then θ_1 is the parameter of interest and (2.6.11) is an orthogonal parameterization. Also,

$$I_{11} \propto \theta_1^{-2} , \quad I_{22} = \theta_1^2/\theta_2^2 , \quad L_{1,1,1} \propto \theta_1^{-3} , \quad L_{112} \propto (\theta_1\theta_2)^{-1} .$$

By (2.5.25), therefore, first order matching is achieved if and only if $\pi(\theta) = d(\theta_2)/\theta_1$. Moreover, by (2.5.26), such a prior is second order matching if and only if $d(\theta_2) \propto \theta_2^{-1}$. Hence $\pi(\theta) \propto (\theta_1\theta_2)^{-1}$ is the unique second order matching prior. This prior becomes proportional to $(\mu_1\mu_2)^{-1}$ when one reverts back to the original $(\mu_1, \mu_2)-$parameterization. Sun (1997) showed that it is also the reference prior when either μ_1 or μ_2 is of interest.   ♣

Example 2.6.7. (Mukerjee and Ghosh, 1997) Consider the $t-$variate normal model with mean vector $\mu = (\mu_1, \ldots, \mu_t)^T$ and dispersion matrix $\gamma^2 \mathcal{I}_t$, where $\mu(\neq 0) \in \mathcal{R}^t$ and $\gamma \ (> 0)$ are unknown, $\mathcal{I}_t$ is the identity matrix of order t, and $t \geq 2$. Interest lies in $\mu^T\mu/\gamma^2$. Reparameterize as

$$\begin{aligned}
\gamma &= g(\theta_1, \theta_2)/\theta_1^{1/2} , \\
\mu_1 &= g(\theta_1, \theta_2) \cos\theta_3 , \\
\mu_2 &= g(\theta_1, \theta_2) \sin\theta_3 \cos\theta_4 ,
\end{aligned}$$

$$\vdots$$

$$\begin{aligned}
\mu_{t-1} &= g(\theta_1, \theta_2) \sin\theta_3 \ldots \sin\theta_t \cos\theta_{t+1} , \\
\mu_t &= g(\theta_1, \theta_2) \sin\theta_3 \ldots \sin\theta_t \sin\theta_{t+1} ,
\end{aligned}$$

where $g(\theta_1, \theta_2) = \theta_2\{\theta_1/(\theta_1 + 2t)\}^{1/2}$ and $\theta_1, \theta_2 > 0$, $0 < \theta_3, \ldots, \theta_t < \pi$, $0 < \theta_{t+1} < 2\pi$, the π appearing in the range of $\theta_3, \ldots, \theta_{t+1}$ being the usual transcendental number (not to be confused with a prior). Then $\theta_1 = \mu^T\mu/\gamma^2$ is the parameter of interest and orthogonal parameterization holds. Furthermore,

$$I_{11} = \frac{1}{2}t\{\theta_1(\theta_1 + 2t)\}^{-1}, \quad I_{22} = (\theta_1 + 2t)/\theta_2^2,$$

$$I_{2s} = L_{11s} = 0 \ (3 \le s \le t+1),$$

$$L_{1,1,1} = -\frac{t(\theta_1 + 3t)}{\theta_1(\theta_1 + 2t)^3}, \quad L_{112} = \frac{t^2}{\theta_1\theta_2(\theta_1 + 2t)^2}.$$

Hence by (2.5.25), a prior $\pi(\theta)$ is first order matching if and only if it is of the form $\pi(\theta) = d(\theta^{(2)})\{\theta_1(\theta_1 + 2t)\}^{-1/2}$, where $\theta^{(2)} = (\theta_2, \ldots, \theta_{t+1})^T$. By (2.5.26), such a prior is second order matching if and only if $d(\theta^{(2)})$ is free from θ_2, i.e., if and only if the prior is of the form

$$\pi(\theta) = d^*(\theta^{(3)})\{\theta_1(\theta_1 + 2t)\}^{-1/2}, \tag{2.6.12}$$

where $d^*(\theta^{(3)})(> 0)$ is any smooth function of $\theta^{(3)} = (\theta_3, \ldots, \theta_{t+1})^T$. In particular, the second order matching prior given by (2.6.12) with

$$d^*(\theta^{(3)}) = \sin^{t-2}\theta_3 \sin^{t-3}\theta_4 \ldots \sin\theta_t \tag{2.6.13}$$

becomes proportional to the nice form $\{\gamma(\mu^T\mu)^{\frac{1}{2}(t-1)}\}^{-1} \ (= \pi^{(1)}(\mu, \gamma)$, say) when one transforms back to the original (μ, γ)−parameterization.

The first order matching prior associated with the choice $d(\theta^{(2)}) = d^*(\theta^{(3)})/\theta_2$, where $d^*(\theta^{(3)})$ is as in (2.6.13), has been of some interest. Under the original (μ, γ)−parameterization, this prior becomes proportional to $\gamma^{-1}(\mu^T\mu + 2t\gamma^2)^{-1/2}(\mu^T\mu)^{-(t-1)/2} \ (= \pi^{(2)}(\mu, \gamma)$, say). It was studied by Datta and Ghosh, J.K. (1995b).

In the setup of this example, now let $t = 1$, so that μ is a scalar. Suppose interest lies in the ratio μ/γ. One may wonder if the priors that would result from taking $t = 1$ in those obtained above have the desired matching property for μ/γ. Satisfyingly, the answer is affirmative. To see this, reparameterize as

$$\theta_1 = \frac{\mu}{\gamma}, \quad \theta_2 = (\mu^2 + 2\gamma^2)^{1/2}.$$

Then θ_1 is the parameter of interest and orthogonal parameterization holds. Furthermore,

$$I_{11} = \frac{2}{2 + \theta_1^2}, \quad I_{22} = \frac{2 + \theta_1^2}{\theta_2^2},$$

$$L_{1,1,1} = -\frac{8\theta_1(3 + \theta_1^2)}{(2 + \theta_1^2)^3}, \quad L_{112} = \frac{4}{\theta_2(2 + \theta_1^2)^2}.$$

Hence by (2.5.25), a prior $\pi(\theta)$ is first order matching if and only if it is of the form $\pi(\theta) = d(\theta_2)(2 + \theta_1^2)^{-1/2}$. If one takes $d(\theta_2) = \theta_2^{-1}$, the corresponding prior in (μ, γ)−parameterization is proportional to $\gamma^{-1}(\mu^2 + 2\gamma^2)^{-1/2}$. Bernardo (1979) derived this as a reference prior for μ/γ. Note that this is identical with the prior $\pi^{(2)}(\mu, \gamma)$ shown above with $t = 1$. On the other hand, the unique second order matching prior obtained by solving (2.5.26) is characterized by $d(\theta_2) = $ constant. In the (μ, γ)−parameterization it is proportional to γ^{-1}. Again, one would get this prior taking $t = 1$ in $\pi^{(1)}(\mu, \gamma)$. ♣

Example 2.6.8. Consider the balanced one-way analysis of variance (ANOVA) setting, with random effects, as given by

$$X_i^{(j)} = \mu + a_i + e_{ij}, \quad 1 \le i \le n, \quad 1 \le j \le t,$$

where the parameter μ represents the general mean, each a_i is univariate normal with mean zero and variance λ_1, each e_{ij} is univariate normal with mean zero and variance λ_2, and the a_i's and the e_{ij}'s are all independent. Here $\mu \ (\in \mathcal{R}^1)$ and $\lambda_1, \lambda_2 \ (> 0)$ are unknown parameters and $t \ (\ge 2)$ is fixed.

For $1 \le i \le n$, let $X_i = (X_i^{(1)}, \ldots, X_i^{(t)})^T$. Then $X_1, \ldots, X_n$ are i.i.d. random variables each following the t−variate normal distribution with mean vector $\mu 1_t$ and dispersion matrix $\lambda_1 E_t + \lambda_2 \mathcal{I}_t$, where 1_t is the $t \times 1$ vector with each element unity, $E_t = 1_t 1_t^T$ and, as in the last example, $\mathcal{I}_t$ is the identity matrix of order t. In the present example, we treat this t−variate normal model as the basic model and consider asymptotics as $n \to \infty$.

Suppose interest lies in the ratio λ_1/λ_2 of the variance components. Reparameterize as

$$\theta_1 = \lambda_1/\lambda_2, \quad \theta_2 = \{\lambda_2^{t-1}(t\lambda_1 + \lambda_2)\}^{1/(2t)}, \quad \theta_3 = \mu, \qquad (2.6.14)$$

where $\theta_1, \theta_2 > 0$ and $\theta_3 \in \mathcal{R}^1$. Then θ_1 is the parameter of interest. With reference to the above t−variate normal model, (2.6.14) is an orthogonal parameterization and

$$I_{11} \propto (1 + t\theta_1)^{-2}, \quad I_{22} \propto \theta_2^{-2}, \quad I_{23} = 0,$$

$$L_{1,1,1} \propto (1 + t\theta_1)^{-3}, \quad L_{112} \propto \{(1 + t\theta_1)^2 \theta_2\}^{-1}, \quad L_{113} = 0.$$

Hence by (2.5.25), first order matching is achieved if and only if $\pi(\theta) = d(\theta^{(2)})/(1 + t\theta_1)$, where $\theta^{(2)} = (\theta_2, \theta_3)^T$. Furthermore, by (2.5.26), such a prior is second order matching if and only if $d(\theta^{(2)})$ is of the form $d^*(\theta_3)/\theta_2$, i.e., if and only if the prior is of the form

$$\pi(\theta) = d^*(\theta_3)/\{(1 + t\theta_1)\theta_2\}, \qquad (2.6.15)$$

where $d^*(\theta_3)(> 0)$ is any smooth function of θ_3. In particular, with $d^*(\theta_3) =$ constant in (2.6.15), one gets the second order matching prior given by $\pi(\theta) \propto \{(1 + t\theta_1)\theta_2\}^{-1}$ that becomes proportional to $\{(t\lambda_1 + \lambda_2)\lambda_2\}^{-1}$ when one reverts back to the original $(\mu, \lambda_1, \lambda_2)$−parameterization. As noted by Berger and Bernardo (1992b) and Ye (1994), this is also a reference prior. It was recommended by Datta and Ghosh, M. (1995b) from other considerations as well.   ♣

2.7 Invariance

In many of the examples considered in the last two sections, the object of interest was a parametric function under an original and often natural parameterization -- e.g., in Example 2.6.8, we were interested in λ_1/λ_2 under the original $(\mu, \lambda_1, \lambda_2)$−parameterization. In order to employ the matching conditions derived earlier in the chapter, the parametric function of interest was reduced to a canonical interest parameter θ_1 via reparameterization. In several examples, the resulting matching priors were also transformed back to their forms under the original parameterization. These steps are justified because of the invariance of the underlying matching problem with respect to one-to-one transformation of the parameter vector. In subsection 2.5.3, we had hinted at this point. The issue of invariance was studied by Datta and Ghosh (1996) and Mukerjee and Ghosh (1997). We follow the latter authors, whose treatment of this topic is somewhat simpler in the present context, to explain such invariance in some detail.

Suppose the model is originally parameterized by $\lambda = (\lambda_1, \ldots, \lambda_p)^T$. Let $g^*(\lambda)$ be a parametric function of interest, $\pi^*(\lambda)$ be a prior on λ, and $f^*(\cdot; \lambda)$ be the common density of the i.i.d. random variables $X_1, X_2, \ldots$. Consider a one-to-one transformation $\lambda = \lambda(\theta)$, where $\theta = (\theta_1, \ldots, \theta_p)^T$, such that the Jacobian matrix $J = ((\partial \lambda_j / \partial \theta_i))$ is nonsingular for all θ. The transformed versions of $g^*(\lambda)$, $\pi^*(\lambda)$ and $f^*(\cdot; \lambda)$ under the θ−parameterization are given respectively by

$$g(\theta) = g^*(\lambda(\theta)) \,, \quad \pi(\theta) = \pi^*(\lambda(\theta))|\det(J)| \,, \quad f(\cdot; \theta) = f^*(\cdot; \lambda(\theta)) \,. \quad (2.7.1)$$

As before, with $X = (X_1, \ldots, X_n)^T$, let $P^\pi\{\cdot | X\}$ be the posterior probability measure under the prior $\pi(\cdot)$ and the θ−parameterization. Similarly, define $P^{\pi^*}\{\cdot | X\}$ with respect to the prior $\pi^*(\cdot)$ and the λ−parameterization. Then for any observational function $\xi(X)$, using (2.7.1),

$$
\begin{aligned}
&P^{\pi^*}\{g^*(\lambda) \leq \xi(X) | X\} \\
&= \frac{\int_{\{\lambda : g^*(\lambda) \leq \xi(X)\}} \pi^*(\lambda)\{\prod_{i=1}^n f^*(X_i; \lambda)\} \mathrm{d}\lambda}{\int \pi^*(\lambda)\{\prod_{i=1}^n f^*(X_i; \lambda)\} \mathrm{d}\lambda} \\
&= \frac{\int_{\{\theta : g^*(\lambda(\theta)) \leq \xi(X)\}} \pi^*(\lambda(\theta))\{\prod_{i=1}^n f^*(X_i; \lambda(\theta))\}|\det(J)| \mathrm{d}\theta}{\int \pi^*(\lambda(\theta))\{\prod_{i=1}^n f^*(X_i; \lambda(\theta))\}|\det(J)| \mathrm{d}\theta} \\
&= \frac{\int_{\{\theta : g(\theta) \leq \xi(X)\}} \pi(\theta)\{\prod_{i=1}^n f(X_i; \theta)\} \mathrm{d}\theta}{\int \pi(\theta)\{\prod_{i=1}^n f(X_i; \theta)\} \mathrm{d}\theta} \\
&= P^\pi\{g(\theta) \leq \xi(X) | X\} \,. \quad\quad (2.7.2)
\end{aligned}
$$

Similarly, with $\lambda = \lambda(\theta)$, from (2.7.1) it is easy to see that

$$P_\lambda\{g^*(\lambda) \leq \xi(X)\} = P_\theta\{g(\theta) \leq \xi(X)\} \,. \quad (2.7.3)$$

In the same way as with a canonical interest parameter, we define a matching prior for a parametric function as one that ensures frequentist validity of the posterior quantiles of the parametric function. If such frequentist validity holds with margin of error $o(n^{-1/2})$ or $o(n^{-1})$ then, as before, the prior is called first or second order matching respectively, for the parametric function.

Since any posterior quantile of a parametric function is an observational function, from (2.7.2) and (2.7.3) it now follows that $\pi^*(\lambda)$ is a matching prior for $g^*(\lambda)$ up to any order of approximation under the $\lambda-$parameterization if and only if $\pi(\theta)$ is so for $g(\theta)$ up to the same order of approximation under the $\theta-$parameterization. But $\pi(\theta)$ and $g(\theta)$ are only transformed versions of $\pi^*(\lambda)$ and $g^*(\lambda)$ respectively under a reparameterization. Hence it is evident that the problem of choosing a matching prior for a parametric function in the sense described above is invariant of the parameterization adopted. In particular, if the gradient vector

$$\bigtriangledown g^*(\lambda) = (\partial g^*(\lambda)/\partial \lambda_1 ,\ldots , \partial g^*(\lambda)/\partial \lambda_p)^T$$

is nonnull for every λ, then in the one-to-one transformation $\lambda \to \theta$, one can always take $\theta_1 = g^*(\lambda)$. This justifies the reduction of the parametric function of interest to a canonical interest parameter through reparameterization as done in many of the examples. Also, when one works with a canonical interest parameter θ_1, the same argument as in (2.7.2) and (2.7.3) entails invariance with respect to the choice of the nuisance parameters $\theta_2,\ldots,\theta_p$. This again justifies the choice of $\theta_2,\ldots,\theta_p$ so as to achieve an orthogonal parameterization.

2.8 General parametric functions and Bayesian tolerance limits

2.8.1 General parametric functions

We continue with the problem of choosing a matching prior for a parametric function of interest. In view of the discussion in the last section, it is perfectly legitimate to reduce the parametric function to a canonical interest parameter via reparameterization and then apply the matching conditions presented earlier in this chapter. In some situations, however, it may be inconvenient to follow this approach. For example, if one wishes to consider several parametric functions with the same model then a separate reparameterization is needed for each such parametric function and this can entail substantial duplication of work since the Fisher information matrix and the quantities like L_{jrs} have to be calculated afresh for each such reparameterization. Moreover, even with a single parametric function of interest, the reparameterization may destroy the natural forms of the information matrix and the L_{jrs} and thus yield matching conditions in the form of awkward partial differential equations that can be hard to solve. From these considerations, it can sometimes help to have a counterpart of Theorem 2.4.1 giving matching conditions directly for a parametric function rather than a canonical interest parameter. Such a result, shown in Theorem 2.8.1 below, helps in handling multiple parametric functions of interest for the same model in a unified manner (see Example 2.8.1 below) in addition to retaining the simplicity that often arises with a natural parameterization (see Example 2.8.2 below). An application to Bayesian tolerance limits with approximate frequentist validity is also discussed later in subsection 2.8.2.

Consider the setup of Section 2.2 with D_j, I^{jr} and L_{jrs} $(1 \leq j, r, s \leq p)$ being as defined there. Let $g(\theta)$ be a smooth parametric function of interest. For $1 \leq j, r \leq p$, let

$$\left.\begin{array}{l} g_j = D_j g(\theta), \quad g_{jr} = D_j D_r g(\theta), \quad b_j = I^{jr} g_r, \\ g_{j0} = g_{jr} b_r, \quad L_{j00} = L_{jrs} b_r b_s. \end{array}\right\} \tag{2.8.1}$$

Also define

$$\Lambda = (I^{jr} g_j g_r)^{1/2}, \quad g_{00} = g_{jr} b_j b_r, \quad L_{000} = L_{jrs} b_j b_r b_s. \tag{2.8.2}$$

The quantities in (2.8.1) and (2.8.2) are all functions of θ and, as usual, the summation convention, with implicit sums ranging from 1 to p, is followed whenever applicable. Assume that the gradient vector $\nabla g(\theta) = (g_1, \ldots, g_p)^T$ is nonnull for every θ. Then $\Lambda > 0$ for every θ and the following result holds.

Theorem 2.8.1. *(a) A prior $\pi(\cdot)$ is first order matching for $g(\theta)$ if and only if it satisfies the partial differential equation*

$$D_j \{\Lambda^{-1} b_j \pi(\theta)\} = 0. \tag{2.8.3}$$

(b) The prior $\pi(\cdot)$ is second order matching for $g(\theta)$ if and only if it satisfies, in addition, the partial differential equation

$$D_r \left[\left\{ \Lambda^{-2} I^{jr} (L_{j00} + 2 g_{j0}) - \Lambda^{-4} b_r \left(\frac{2}{3} L_{000} + 2 g_{00} \right) \right\} \pi(\theta) \right]$$

$$- D_j D_r \{\Lambda^{-2} b_j b_r \pi(\theta)\} = 0. \tag{2.8.4}$$

♣

In particular, if $g(\theta) = \theta_1$, then $g_1 = 1$, $g_2 = \ldots = g_p = 0$, $\Lambda = (I^{11})^{1/2}$, $g_{00} = 0$, $b_j = I^{j1}$ and $g_{j0} = 0$ $(1 \leq j \leq p)$. Hence it is not hard to verify that then (2.8.3) and (2.8.4) reduce respectively to the corresponding matching conditions (2.4.11) and (2.4.12) shown in Theorem 2.4.1. Thus Theorem 2.8.1 is in agreement with Theorem 2.4.1. In principle, Theorem 2.8.1 can be obtained from Theorem 2.4.1 using the invariance argument discussed in the last section. For this purpose, one has to reparameterize $\theta \to \theta^*$, where $\theta^* = (\theta_1^*, \ldots, \theta_p^*)^T$ and $\theta_1^* = g(\theta)$, then express the matching conditions of Theorem 2.4.1 in terms of the θ^*-parameterization, and finally revert back to the θ-parameterization. The algebra underlying these steps, akin to that in Section 3 of Datta and Ghosh (1996), is, however, quite challenging. Mukerjee and Reid (2001) outlined a proof of Theorem 2.8.1 from first principles. We omit details here. Incidentally, the first order matching condition (2.8.3) was obtained also by Stein (1985) and Datta and Ghosh, J.K. (1995b) in other contexts. The work of the latter authors will be reviewed in the next chapter.

Example 2.8.1. (Mukerjee and Reid, 2001) Consider the bivariate normal model with means θ_1, θ_2 ($\in \mathcal{R}^1$), standard deviations θ_3, θ_4 (> 0) and correlation coefficient θ_5, with $|\theta_5| < 1$. Then

$$I^{11} = \theta_3^2 , \quad I^{22} = \theta_4^2 , \quad I^{33} = \frac{1}{2}\theta_3^2 , \quad I^{44} = \frac{1}{2}\theta_4^2 , \quad I^{55} = (1 - \theta_5^2)^2 ,$$

$$I^{12} = \theta_3\theta_4\theta_5 , \quad I^{34} = \frac{1}{2}\theta_3\theta_4\theta_5^2 ,$$

$$I^{35} = \frac{1}{2}\theta_3\theta_5(1 - \theta_5^2) , \quad I^{45} = \frac{1}{2}\theta_4\theta_5(1 - \theta_5^2) ,$$

$$I^{13} = I^{14} = I^{15} = I^{23} = I^{24} = I^{25} = 0 ,$$

$$L_{122} = L_{133} = L_{134} = L_{144} = L_{145} = L_{155} = 0 ,$$

$$L_{222} = L_{223} = L_{233} = L_{234} = L_{235} = L_{244} = L_{245} = L_{255} = 0 ,$$

$$L_{124} = -\frac{\theta_5}{\theta_3\theta_4^2(1 - \theta_5^2)} , \quad L_{125} = \frac{1 + \theta_5^2}{\theta_3\theta_4(1 - \theta_5^2)^2} ,$$

$$L_{224} = \frac{2}{\theta_4^3(1 - \theta_5^2)} , \quad L_{225} = -\frac{2\theta_5}{\theta_4^2(1 - \theta_5^2)^2} ,$$

$$L_{345} = \frac{\theta_5(1 + \theta_5^2)}{\theta_3\theta_4(1 - \theta_5^2)^2} , \quad L_{555} = -\frac{4\theta_5(3 + \theta_5^2)}{(1 - \theta_5^2)^3} ,$$

$$\theta_3^3 L_{333} = \theta_4^3 L_{444} = \frac{10 - 4\theta_5^2}{1 - \theta_5^2} , \quad \theta_3 L_{334} = \theta_4 L_{344} = -\frac{2\theta_5^2}{\theta_3\theta_4(1 - \theta_5^2)} ,$$

$$\theta_3^2 L_{335} = \theta_4^2 L_{445} = -\frac{2\theta_5(2 - \theta_5^2)}{(1 - \theta_5^2)^2} , \quad \theta_3 L_{355} = \theta_4 L_{455} = \frac{2(1 + \theta_5^2)}{(1 - \theta_5^2)^2} .$$

The remaining L_{jrs} are not needed in this example.

We consider three parametric functions of interest, namely,

(a) $g(\theta) = \theta_3/\theta_4$,
(b) $g(\theta) = \theta_3^2\theta_4^2(1 - \theta_5^2)$,
(c) $g(\theta) = \theta_2 + (\theta_4\theta_5/\theta_3)(k - \theta_1)$.

In (c), k is a constant which does not involve θ. The choices of $g(\theta)$ in (a), (b) and (c) correspond to the ratio of standard deviations, generalized variance and regression function respectively. We shall consider a natural class of priors of the form

$$\pi(\theta) = \{\theta_3^{s_3}\theta_4^{s_4}(1 - \theta_5^2)^{s_5}\}^{-1} , \tag{2.8.5}$$

where s_3, s_4 and s_5 are real numbers.

First, let $g(\theta)$ be as in (a). Then from (2.8.1), (2.8.2) and the expressions for the I^{jr} and L_{jrs} as shown above,

$$b_1 = b_2 = b_5 = 0, \quad \theta_3^{-1}b_3 = -\theta_4^{-1}b_4 = \frac{1}{2}(\theta_3/\theta_4)(1-\theta_5^2),$$

$$\Lambda = (\theta_3/\theta_4)(1-\theta_5^2)^{1/2}, \quad L_{000} = L_{100} = L_{200} = g_{10} = g_{20} = g_{50} = 0,$$

$$\theta_3 L_{300} = \theta_4 L_{400} = \frac{1}{2}(\theta_3/\theta_4)^2(1-\theta_5^2)(5-\theta_5^2),$$

$$L_{500} = -\frac{1}{2}(\theta_3/\theta_4)^2\theta_5(5-\theta_5^2), \quad g_{00} = (\theta_3/\theta_4)^3(1-\theta_5^2)^2,$$

$$g_{30} = \frac{1}{2}(\theta_3/\theta_4^2)(1-\theta_5^2), \quad g_{40} = -\frac{3}{2}(\theta_3^2/\theta_4^3)(1-\theta_5^2).$$

Thus a prior of the form (2.8.5) satisfies (2.8.3) and is hence first order matching if and only if $s_3 = s_4$. Furthermore, such a prior also satisfies (2.8.4) and is hence second order matching if and only if $s_3 = s_4 = 0$, $s_5 = 1$.

In a similar manner, with $g(\theta)$ as in (b), first order matching holds if and only if $s_3 + s_4 = 2$ and all such priors are second order matching as well. Finally, with $g(\theta)$ as in (c), one gets first order matching if and only if $s_4 = 2(s_5 - 1)$ and second order matching if and only if $s_4 = 0$, $s_5 = 1$. Thus a prior of the form (2.8.5) is first order matching for all the three parametric functions if and only if $s_3 = s_4 = 1$, $s_5 = 3/2$. On the other hand, no prior of this kind is second order matching for all three of them.

This example underscores an advantage of using Theorem 2.8.1. Even though three parametric functions are considered here, one needs to compute the quantities I^{jr} and L_{jrs} once and for all.   ♣

Example 2.8.2. (Sun and Ye, 1995) This concerns the product of p (≥ 2) independent normal means. Let

$$f(x;\theta) = \prod_{j=1}^{p} \phi(x^{(j)} - \theta_j),$$

where $x = (x^{(1)}, \ldots, x^{(p)})^T \in \mathcal{R}^p$ and $\theta_1, \ldots, \theta_p > 0$. The per observation Fisher information matrix equals the identity matrix of order p. Suppose interest lies in the product $g(\theta) = \prod_{j=1}^{p} \theta_j$. By (2.8.1) and (2.8.2) then $b_j = g(\theta)/\theta_j$ $(1 \leq j \leq p)$ and $\Lambda = g(\theta)(\sum_{j=1}^{p} \theta_j^{-2})^{1/2}$. Hence by (2.8.3),

$$\pi(\theta) \propto (\prod_{j=1}^{p} \theta_j)(\sum_{j=1}^{p} \theta_j^{-2})^{1/2} \tag{2.8.6}$$

is a first order matching prior for $g(\theta)$. In particular, for $p = 2$, the above is in agreement with the findings of Example 2.6.2 and, as seen there, no second order matching prior exists for $g(\theta)$. Sun and Ye (1995) noted that (2.8.6) is also a reference prior. They obtained (2.8.6) via a reparameterization that reduces $g(\theta)$ to a canonical interest parameter. The present derivation is simpler since it exploits, through the use of Theorem 2.8.1, the simple form of the information matrix under the natural parameterization.   ♣

From the discussion in the beginning of this subsection, it is clear that there are two possible routes for obtaining a matching prior for a parametric function of interest. One can either use the matching conditions of Theorem 2.4.1 (or simplified versions thereof as obtained in Section 2.5 for specific situations) via reduction to a canonical interest parameter or employ those of Theorem 2.8.1 to handle the parametric function directly. The invariance argument of Section 2.7 implies the equivalence of the two approaches and a choice between them depends on convenience. Many of the examples in Sections 2.5 and 2.6 illustrate applications of the first approach while the last two examples apply the second approach to particular problems. Quite a few other applications of these matching conditions have been reported in the literature. This work is briefly indicated below.

Applications to more general versions of the Fieller–Creasy problem (cf. Example 2.6.1) were studied by Yin and Ghosh (2001) and Ghosh, Rousseau and Kim (2001). The former authors considered general location-scale models and the latter authors studied the problem in the bivariate normal case. In the same spirit, Ghosh, Yin and Kim (2003) reported results pertaining to a ratio of two regression coefficients in the multiple linear regression model. Sun and Ye (1999) extended the findings of Examples 2.6.2 and 2.8.2, concerning the product of independent normal means, to the case of unknown population standard deviations. Various applications to the balanced one-way random ANOVA model (cf. Example 2.6.8) were investigated by Datta and Ghosh, M. (1995a, b), Chung and Dey (1998) and Kim, Kang and Lee (2001). An extension to the unbalanced case was studied by Datta, Ghosh and Kim (2002). Further results on three-stage nested designs were reported by Kim, Kang and Lee (2003). Berger, Philippe and Robert (1998) considered an application where the parametric function of interest is a quadratic function of bivariate normal means. Other applications include those to group invariance models (Datta and Ghosh, J. K., 1995a), univariate and multivariate calibration problems (Ghosh, Carlin and Srivastava, 1995; Yin, 2000; Eno and Ye, 2001), stress-strength systems in the simple exponential and Weibull cases (Lee, 1998; Sun, Ghosh and Basu, 1998; Kim, Chang and Kang, 2001), bivariate competing risk models (Wang and Ghosh, 2000), and linear models with intraclass correlation structure for error (Ghosh and Heo, 2003). We refer to the original papers for detailed accounts of these applications often along with discussion on possible connection with reference priors.

2.8.2 Bayesian tolerance limits

In the setup of Section 2.2, suppose the i.i.d. random variables $\{X_i\}$, $i \geq 1$, are one-dimensional and let $F(x; \theta)$ denote their common cumulative distribution function, where θ continues to be possibly multidimensional. Consider a Bayesian tolerance interval of the form (T_0, ∞) such that the relation

$$P^\pi\{1 - F(T_0; \theta) \geq \beta | X\} = \alpha \qquad (2.8.7)$$

holds, where α and β $(0 < \alpha, \beta < 1)$ are preassigned and $X = (X_1, \ldots, X_n)^T$. Clearly, the tolerance limit T_0 depends on both X and the prior $\pi(\cdot)$ in addition to α and β. By (2.8.7), the interval (T_0, ∞) covers at least a proportion β of the population represented by $F(x; \theta)$ with posterior probability α. In this sense, following Guttman (1970, Ch. 9), it is called a β-content tolerance interval with posterior credibility level α. The interval has frequentist validity with margin of error $o(n^{-r/2})$ $(r = 1$ or $2)$ provided

$$P_\theta\{1 - F(T_0; \theta) \geq \beta\} = \alpha + o(n^{-r/2}) . \qquad (2.8.8)$$

If the underlying prior $\pi(\cdot)$ ensures (2.8.8) irrespective of the choice of α and β, then it is called first or second order matching for tolerance limits depending on whether r equals 1 or 2 respectively.

Observe that

$$1 - F(T_0; \theta) \geq \beta \iff g_\beta(\theta) \geq T_0 ,$$

where $g_\beta(\theta)$ is the $(1 - \beta)$th quantile of the population given by $F(x; \theta)$. Hence by (2.8.7), T_0 can be interpreted as the $(1 - \alpha)$th posterior quantile of $g_\beta(\theta)$. Consequently, as noted in Mukerjee and Reid (2001), a prior is first or second order matching for tolerance limits if and only if it is a matching prior for $g_\beta(\theta)$, upto the same order of approximation, whatever be the choice of β. Therefore, one can use Theorem 2.8.1, with $g(\theta) = g_\beta(\theta)$, to characterize such matching priors for tolerance limits. The following example illustrates this point.

Example 2.8.3. (Mukerjee and Reid, 2001) We revisit the location-scale model considered in (2.5.16). Then $g_\beta(\theta) = \theta_1 + q_\beta \theta_2$, where the constant q_β is the $(1 - \beta)$th quantile of the population represented by the density $f^*(\cdot)$. Following (2.5.17),

$$I^{jr} = u^{jr} \theta_2^2 \quad (j, r = 1, 2) , \qquad (2.8.9)$$

the u^{jr} being constants free from θ. With $g(\theta) = g_\beta(\theta) = \theta_1 + q_\beta \theta_2$, by (2.8.1), (2.8.2) and (2.8.9),

$$b_j = (u^{j1} + u^{j2}q_\beta)\theta_2^2 \quad (j = 1, 2) , \quad \Lambda \propto \theta_2 .$$

Therefore, the first order matching condition (2.8.3) becomes

$$D_j\{(u^{j1} + u^{j2}q_\beta)\theta_2 \pi(\theta)\} = 0 ,$$

which holds for every β if and only if

$$u^{j1}D_j\{\theta_2 \pi(\theta)\} = 0 , \quad u^{j2}D_j\{\theta_2 \pi(\theta)\} = 0 ,$$

i.e., if and only if

$$D_1\{\theta_2\pi(\theta)\} = 0 , \quad D_2\{\theta_2\pi(\theta)\} = 0 , \qquad (2.8.10)$$

noting that the matrix $((u^{jr}))$ is positive definite in view of (2.8.9). Thus one gets a unique first order matching prior for tolerance limits, namely, $\pi(\theta) \propto \theta_2^{-1}$, which uniquely satisfies (2.8.10). As seen in (2.5.17), $L_{jrs} \propto \theta_2^{-3}$ for the present location-scale model. Hence, using (2.8.4), one can check that the prior $\pi(\theta) \propto \theta_2^{-1}$ is second order matching as well for tolerance limits. This reinforces the findings in subsection 2.5.2 where the same prior was seen to enjoy the second order matching property when either θ_1 or θ_2 is the parameter of interest.   ♣

2.9 Matching alternative coverage probabilities

Let $g(\theta)$ be a smooth parametric function of interest and $g^{(1-\alpha)}(\pi, X)$ be the $(1 - \alpha)$th posterior quantile of $g(\theta)$ given $X = (X_1, \ldots, X_n)^T$, under a prior $\pi(\cdot)$. If $\pi(\cdot)$ is first or second order matching for $g(\theta)$ then Bayesian credible sets of the form $(-\infty, g^{(1-\alpha)}(\pi, X)]$ for $g(\theta)$ have correct frequentist coverage as well, with margin of error $o(n^{-1/2})$ or $o(n^{-1})$ respectively. In this sense, such Bayesian credible sets can also be interpreted as frequentist confidence sets. From the frequentist point of view, however, the probability for a confidence set to include an alternative value of the parametric function of interest is as important as that of the true coverage (Lehmann, 1986, Ch. 3). Such an alternative coverage probability indicates how selective a confidence set is. In the present context, considering contiguous alternatives with respect to an information based Riemannian metric (Amari, 1985, Ch. 6), one may thus wish to examine how far a first or second order matching prior for $g(\theta)$ also matches

$$P_\theta[g(\theta) + \delta\{(\nabla g)^T I^{-1}(\nabla g)/n\}^{1/2} \le g^{(1-\alpha)}(\pi, X)]$$

with the corresponding posterior probability, up to the same order of approximation and for each δ and α. As before, in the above, I is the per observation Fisher information matrix and ∇g is the gradient vector of $g(\theta)$. Also, the scalar δ is free from n, θ and X; the cases $\delta = 0$ and $\delta \ne 0$ correspond to the true and alternative coverage probabilities respectively.

Mukerjee and Reid (1999a) argued that if a matching prior for $g(\theta)$ also matches the alternative coverage probabilities in the above sense, then there is a stronger justification for calling it noninformative in so far as agreement with a frequentist is concerned. They worked with a canonical interest parameter θ_1, i.e., took $g(\theta) = \theta_1$, so that $\nabla g = (1, 0, \ldots, 0)^T$, $(\nabla g)^T I^{-1}(\nabla g) = I^{11}$, and $g^{(1-\alpha)}(\pi, X)$ reduces to $\theta_1^{(1-\alpha)}(\pi, X)$ as introduced in Section 2.1. Accordingly, they characterized priors $\pi(\cdot)$ which match true as well as alternative coverage probabilities in the sense that

$$P_\theta\{\theta_1 + \delta(I^{11}/n)^{1/2} \leq \theta_1^{(1-\alpha)}(\pi, X)\}$$
$$= E_\theta[P^\pi\{\theta_1 + \delta(I^{11}/n)^{1/2} \leq \theta_1^{(1-\alpha)}(\pi, X)|X\}] + o(n^{-r/2}) \quad (2.9.1)$$

for each δ and α, where $r = 1$ or 2. It was shown by Mukerjee and Reid (1999a) that (2.9.1) holds with $r = 1$, for each δ and α if and only if

$$D_j\{\pi(\theta)I^{j1}(I^{11})^{-1/2}\} = 0. \quad (2.9.2)$$

They also showed that (2.9.1) holds with $r = 2$, for each δ and α if and only if, in addition, the conditions

$$D_u\{\pi(\theta)L_{jrs}\tau^{jr}\sigma^{su}\} - D_jD_r\{\pi(\theta)\tau^{jr}\} = 0, \quad (2.9.3)$$

$$D_u\{\pi(\theta)L_{jrs}\tau^{jr}\tau^{su}\} = 0, \quad (2.9.4)$$

$$D_j[\pi(\theta)I^{jr}(I^{11})^{-1/2}\{D_r(I^{11})^{1/2}\}] = 0, \quad (2.9.5)$$

$$D_j[\pi(\theta)\tau^{jr}(I^{11})^{-1/2}\{D_r(I^{11})^{1/2}\}] = 0, \quad (2.9.6)$$

are met, the notation in the above being as introduced in Section 2.2. The derivation of these conditions involves use of the shrinkage argument and we refer to Mukerjee and Reid (1999a) for the details.

Interestingly, (2.9.2) is identical with the first order matching condition (2.4.11) given in Theorem 2.4.1. Hence we reach the satisfying conclusion that first order matching priors for θ_1 also match the alternative coverage probabilities up to the same order of approximation. In particular, Jeffreys' prior enjoys this property in the absence of nuisance parameters.

Next observe that (2.9.3) and (2.9.4) together imply the condition (2.4.12) of Theorem 2.4.1 and that the conditions (2.9.2)–(2.9.6) are more stringent than (2.4.11) and (2.4.12). Thus a second order matching prior for θ_1 may or may not match the alternative coverage probabilities up to the same order of approximation. This opens up the possibility of further discrimination among the second order matching priors on the basis of alternative coverage probabilities, a point that has been illustrated in Example 2.9.1 below.

The special case of orthogonal parameterization is of interest. As in subsection 2.5.3, then a prior satisfies (2.9.2) if and only if it is of the form $\pi(\theta) = d(\theta^{(2)})I_{11}^{1/2}$, where $d(\cdot)(> 0)$ is any smooth function of $\theta^{(2)} = (\theta_2, \ldots, \theta_p)^T$. Using the standard regularity condition (cf. (2.5.4))

$$D_r(I_{11}) = -(L_{r,11} + L_{r11}),$$

it is not hard to see that such a prior satisfies (2.9.3)–(2.9.6) if and only if

$$\sum_{s=2}^{p}\sum_{u=2}^{p} D_u\{d(\theta^{(2)})I_{11}^{-1/2}I^{su}L_{11s}\} = 0, \quad (2.9.7)$$

$$\sum_{s=2}^{p} \sum_{u=2}^{p} D_u \{d(\theta^{(2)}) I_{11}^{-1/2} I^{su} L_{s,11}\} = 0 , \qquad (2.9.8)$$

$$D_1(I_{11}^{-3/2} L_{111}) = 0 , \quad D_1(I_{11}^{-3/2} L_{1,11}) = 0 . \qquad (2.9.9)$$

Note that the two conditions in (2.9.9) are model conditions.

Example 2.9.1. (Mukerjee and Reid, 1999a) This concerns a bivariate normal regression coefficient and is in continuation of Example 2.5.3. We work with the $\theta-$parameterization shown in (2.5.27). As noted earlier, this is an orthogonal parameterization. Furthermore,

$$I_{11} = \theta_3/\theta_2 , \quad I_{22} = \frac{1}{2}\theta_2^{-2} , \quad I_{33} = \frac{1}{2}\theta_3^{-2} ,$$

$$I_{23} = I_{24} = I_{25} = I_{34} = I_{35} = 0 ,$$

$$L_{111} = L_{113} = L_{114} = L_{115} = 0 ,$$

$$L_{1,11} = L_{2,11} = L_{4,11} = L_{5,11} = 0 ,$$

$$L_{112} = \theta_3/\theta_2^2 , \quad L_{3,11} = -\theta_2^{-1} .$$

It was seen in Example 2.5.3 that a prior $\pi(\cdot)$ is second order matching for θ_1 if and only if it is of the form $\pi(\theta) = d(\theta^{(2)}) I_{11}^{1/2}$, where $d(\theta^{(2)})$ has the structure

$$d(\theta^{(2)}) = \tilde{d}(\theta^{(3)})/\theta_2^{1/2} , \qquad (2.9.10)$$

$\tilde{d}(\theta^{(3)})(> 0)$ being any smooth function of $\theta^{(3)} = (\theta_3, \theta_4, \theta_5)^T$.

Clearly, the model conditions in (2.9.9) are met here. One can also check that any $d(\theta^{(2)})$ as in (2.9.10) satisfies (2.9.7). Moreover, any such $d(\theta^{(2)})$ satisfies (2.9.8) if and only if $\tilde{d}(\theta^{(3)}) = d^*(\theta_4, \theta_5)/\theta_3^{3/2}$, i.e., if and only if

$$d(\theta^{(2)}) = d^*(\theta_4, \theta_5)/(\theta_2^{1/2} \theta_3^{3/2}) , \qquad (2.9.11)$$

where $d^*(\cdot)(> 0)$ is any smooth function of θ_4 and θ_5. Since $I_{11} = \theta_3/\theta_2$, it follows that a second order matching prior for θ_1 also matches the alternative coverage probabilities up to the same order of approximation if and only if it is of the form $\pi(\theta) = d^*(\theta_4, \theta_5)/(\theta_2\theta_3)$. This, indeed, narrows down the class (2.5.29) of second order matching priors.   ♣

2.10 Propriety of posteriors

The matching priors obtained in the specific cases and examples considered in this chapter are almost invariably improper (rare exceptions are first order matching priors in Example 2.6.1 with suitably chosen $d(\theta_2)$, like $d(\theta_2) = e^{-\theta_2}$). As noted in Section 2.2, for any such improper prior $\pi(\cdot)$, a basic requirement is that it must ensure the propriety of the posterior, i.e., guarantee

$$\int \pi(\theta)\{\prod_{i=1}^{n} f(X_i;\theta)\}d\theta < \infty ,\qquad (2.10.1)$$

with P_θ−probability unity for all θ, whenever n is sufficiently large. In (2.10.1), the integral is over the parameter space and, as usual, $f(\cdot;\theta)$ is the common density of the i.i.d. random variables $\{X_i\}$, $i \geq 1$.

The matching priors $\pi(\theta) = $ constant or $\pi(\theta) \propto \theta^{-1}$ obtained in subsection 2.5.1 for the one-parameter location or scale models as well as the matching prior $\pi(\theta) \propto \theta_2^{-1}$ obtained in subsection 2.5.2 and Example 2.8.3 for the location-scale model satisfy the above requirement. This can be readily checked, for example, with the normal or Cauchy location models, the normal, Cauchy or gamma scale models, and the normal or Cauchy location-scale models.

The same satisfying conclusion holds for the other examples too. One can verify that the following matching priors considered in the examples entail propriety of the posteriors, with P_θ−probability unity for all θ, whenever n is sufficiently large:

(a) Jeffreys' prior in many situations covered by Example 2.5.1 (e.g., in the important special case $t = 1$, $\rho_1(\theta) = \theta$, with $\theta \in \mathcal{R}^1$);
(b) Jeffreys' prior in Example 2.5.2;
(c) the second order matching priors considered in Examples 2.5.3, 2.6.4, 2.6.7, 2.6.8 and 2.9.1 with $d^*(\theta^{(3)}) = \theta_3^{-s}$ (s real), $d^*(\theta_2) = $ constant, $d^*(\theta^{(3)})$ as in (2.6.13), $d^*(\theta_3) = $ constant, and $d^*(\theta_4,\theta_5) = $ constant, respectively;
(d) the unique second order matching priors obtained in Examples 2.5.4, 2.6.1, 2.6.5 and 2.6.6 as well as the second order matching prior reported in Example 2.6.3;
(e) the reference prior in Example 2.5.5;
(f) the first order matching prior for θ_1 in Example 2.5.6 with $d(\theta_2) = \theta_2^{-1}$; also the unique second order matching prior for θ_2 in the same example;
(g) the priors reported in Example 2.5.7;
(h) the first order matching priors given by $d(\theta_2) = $ constant in Example 2.6.2, and (2.8.6) in Example 2.8.2;
(i) all the matching priors considered in Example 2.8.1

For illustration, Examples 2.6.5 and 2.6.8 are revisited below. We also refer to Liseo (1993) and Garvan and Ghosh (1999) in connection with Examples 2.5.5 and 2.5.6, and to Garvan and Ghosh (1999), Sun (1997) and Sun and Ye (1999) in connection with Examples 2.5.7, 2.6.6 and 2.8.2 respectively.

Example 2.6.5 (revisited) For $1 \leq i \leq n$, let $X_i = (X_i^{(1)},\ldots,X_i^{(t)})^T$. Define $Y_j = \sum_{i=1}^{n} X_i^{(j)}(1 \leq j \leq t)$. Note that $Y_1,\ldots,Y_t > 0$, with P_θ−probability unity for all θ. Since $z_1 + \ldots + z_t = 0$, we have

$$\prod_{i=1}^{n} f(X_i; \theta) = \theta_2^{-nt} \exp\left(-\theta_2^{-1} \sum_{j=1}^{t} Y_j e^{-\theta_1 z_j}\right).$$

Hence, with the unique second order matching prior $\pi(\theta) \propto \theta_2^{-1}$, following (2.10.1), we consider the integral

$$\int_{-\infty}^{\infty} \int_{0}^{\infty} \theta_2^{-(nt+1)} \exp\left(-\theta_2^{-1} \sum_{j=1}^{t} Y_j e^{-\theta_1 z_j}\right) d\theta_2 d\theta_1$$

$$= \Gamma(nt) \int_{-\infty}^{\infty} \left(\sum_{j=1}^{t} Y_j e^{-\theta_1 z_j}\right)^{-nt} d\theta_1,$$
(2.10.2)

where, as before, $\Gamma(\cdot)$ is the gamma function. Since $z_1 + \ldots + z_t = 0$ and $z_1, \ldots, z_t$ are not all equal, without loss of generality, let $z_1 < 0$ and $z_t > 0$. Then

$$\int_{0}^{\infty} \left(\sum_{j=1}^{t} Y_j e^{-\theta_1 z_j}\right)^{-nt} d\theta_1 \leq \int_{0}^{\infty} (Y_1 e^{-\theta_1 z_1})^{-nt} d\theta_1 < \infty$$

and

$$\int_{-\infty}^{0} \left(\sum_{j=1}^{t} Y_j e^{-\theta_1 z_j}\right)^{-nt} d\theta_1 \leq \int_{-\infty}^{0} (Y_t e^{-\theta_1 z_t})^{-nt} d\theta_1 < \infty.$$

Hence by (2.10.2), the prior $\pi(\theta) \propto \theta_2^{-1}$ ensures propriety of the posterior in exponential regression model with P_θ–probability unity for all θ. This happens for every $n \geq 1$. ♣

Example 2.6.8 (revisited) For $1 \leq i \leq n$, let $X_i = (X_i^{(1)}, \ldots, X_i^{(t)})^T$ and $\overline{X}_i = t^{-1} \sum_{j=1}^{t} X_i^{(j)}$. Define

$$\overline{X} = (nt)^{-1} \sum_{i=1}^{n} \sum_{j=1}^{t} X_i^{(j)}, \quad Y_1 = t \sum_{i=1}^{n} (\overline{X}_i - \overline{X})^2, \quad Y_2 = \sum_{i=1}^{n} \sum_{j=1}^{t} (X_i^{(j)} - \overline{X}_i)^2.$$

Let $n \geq 2$. Assume that Y_1 and Y_2 are both positive – for $n \geq 2$ (recall that $t \geq 2$), this holds with P_θ–probability unity for all θ. Consider the second order matching prior (2.6.15) with $d^*(\theta_3) = $ constant. As noted earlier, this prior becomes proportional to $\{(t\lambda_1 + \lambda_2)\lambda_2\}^{-1}$ when one reverts back to the original $(\mu, \lambda_1, \lambda_2)$–parameterization. Hence, following (2.10.1), it is enough to consider the integral

$$\int_{0}^{\infty} \int_{0}^{\infty} \int_{-\infty}^{\infty} \{(t\lambda_1 + \lambda_2)\lambda_2\}^{-1} \prod_{i=1}^{n} \left[\{(2\pi)^{t/2}(t\lambda_1 + \lambda_2)^{1/2} \lambda_2^{(t-1)/2}\}^{-1}\right.$$

$$\times \exp\left\{ -\frac{1}{2}\left(\frac{t(\overline{X}_i - \mu)^2}{t\lambda_1 + \lambda_2} + \frac{1}{\lambda_2}\sum_{j=1}^{t}(X_i^{(j)} - \overline{X}_i)^2 \right) \right\} \Big] d\mu d\lambda_1 d\lambda_2$$

$$= (2\pi)^{-nt/2}\int_0^\infty \int_0^\infty \int_{-\infty}^\infty \left\{ (t\lambda_1 + \lambda_2)^{\frac{1}{2}n+1}\lambda_2^{\frac{1}{2}n(t-1)+1} \right\}^{-1}$$

$$\times \exp\left\{ -\frac{1}{2}\left(\frac{Y_1 + nt(\mu - \overline{X})^2}{t\lambda_1 + \lambda_2} + \frac{Y_2}{\lambda_2} \right) \right\} d\mu d\lambda_1 d\lambda_2 . \qquad (2.10.3)$$

In (2.10.3), π is the usual transcendental number (not to be confused with a prior). First integrating μ out, the integral in (2.10.3) exists finitely if and only if

$$\int_0^\infty \int_0^\infty \left\{ (t\lambda_1 + \lambda_2)^{\frac{1}{2}(n+1)}\lambda_2^{\frac{1}{2}n(t-1)+1} \right\}^{-1} \exp\left\{ -\frac{1}{2}\left(\frac{Y_1}{t\lambda_1 + \lambda_2} + \frac{Y_2}{\lambda_2} \right) \right\} d\lambda_1 d\lambda_2$$

is finite. Transforming $u = (t\lambda_1 + \lambda_2)^{-1}$, $v = \lambda_2^{-1}$, the integral given above becomes

$$t^{-1}\int_0^\infty \int_u^\infty u^{\frac{1}{2}(n-3)}v^{\frac{1}{2}n(t-1)-1} \exp\left\{ -\frac{1}{2}\left(uY_1 + vY_2 \right) \right\} dv du ,$$

and, since $t \geq 2$ and $n \geq 2$, the above exists finitely. Thus propriety of the posterior in variance components model holds, with P_θ−probability unity for all θ, whenever $n \geq 2$.

3

Matching Priors for Distribution Functions

3.1 Introduction

Matching priors for posterior quantiles were discussed at length in the previous chapter. These priors concern a single parameter (see Theorem 2.4.1) or a single parametric function (see Theorem 2.8.1) of interest. Since quantiles are intimately linked with the cumulative distribution function (c.d.f.), one may wonder how far these results carry through when matching is done via c.d.f. instead of quantiles. Continuing with a one-dimensional parameter or parametric function of interest, this issue is addressed in Section 3.2. The results in this section lead to the satisfying conclusion that first order matching priors for quantiles remain so when the analysis is based on a comparison of the posterior and frequentist c.d.f.'s.

We next turn to the situation where interest lies in several parameters or parametric functions. Then posterior quantiles are not well-defined but the joint posterior c.d.f. remains meaningful and provides a viable route for finding matching priors. In fact, consideration of c.d.f. has the potential of yielding a single matching prior even for multiple parameters or parametric functions of interest. These aspects are explored in Section 3.3 from various standpoints.

3.2 C.d.f. matching priors for a single parametric function

3.2.1 Scalar interest parameter

In this chapter, the setup and assumptions, on both the model and the prior $\pi(\cdot)$, are as in Section 2.2. Unless otherwise specified, the notation is also as described there. Suppose first that θ_1 is the one-dimensional parameter of interest. Let $y = (n/c^{11})^{1/2}(\theta_1 - \hat{\theta}_1)$, which is an approximate standardized version of θ_1 in the posterior setup. From (2.3.18), recall that an expansion for the posterior density of y, under $\pi(\cdot)$, is given by

$$\pi_y(y|X) = \phi(y)[1 + n^{-1/2}\{G_1(\pi)J_1(y) + G_3 J_3(y)\}$$
$$+ n^{-1}\{G_2(\pi)J_2(y) + G_4(\pi)J_4(y) + G_6 J_6(y)\}] + o(n^{-1}),$$

where $J_i(\cdot)$ is the Hermite polynomial of degree i (cf. (1.3.10)), and $G_1(\pi)$, G_3 etc. are as shown in (2.3.9)–(2.3.13). Hence for any w, free from n, θ and X,

$$P^{\pi}\{y \leq w|X\} = \Phi(w) - n^{-1/2}\phi(w)\{G_1(\pi) + G_3 J_2(w)\}$$
$$- n^{-1}\phi(w)\{G_2(\pi)J_1(w) + G_4(\pi)J_3(w) + G_6 J_5(w)\}$$
$$+ o(n^{-1}), \tag{3.2.1}$$

where $P^{\pi}\{\cdot|X\}$ is the posterior probability measure under the prior $\pi(\cdot)$. If one proceeds along the line of Step 2 of Section 2.4 and employs (2.2.1)–(2.2.6) and (2.3.4)–(2.3.13), then after some algebra (3.2.1) yields

$$E_{\theta}[P^{\pi}\{y \leq w|X\}]$$
$$= \Phi(w) - n^{-1/2}\phi(w)[Q_1(w,\theta) + (I^{11})^{-1/2}I^{j1}\{\pi_j(\theta)/\pi(\theta)\}]$$
$$- n^{-1}\phi(w)\Big[Q_2(w,\theta) + \frac{1}{6}(w^3 + 3w)L_{jrs}\tau^{jr}\tau^{sv}\{\pi_v(\theta)/\pi(\theta)\}$$
$$+ \frac{1}{2}w\{\pi(\theta)\}^{-1}\tau^{jr}\{\pi_{jr}(\theta) + L_{jrs}\sigma^{sv}\pi_v(\theta) + L_{jsv}\sigma^{sv}\pi_r(\theta)\}\Big]$$
$$+ o(n^{-1}), \tag{3.2.2}$$

for all θ, where $Q_1(w,\theta)$ and $Q_2(w,\theta)$ are at most of order $O(1)$ and do not involve $\pi(\cdot)$ or its derivatives. The explicit forms of these quantities are not needed in the sequel. As in Step 3 of Section 2.4, considering a counterpart of (3.2.2) under an auxiliary prior, it follows that

$$P_{\theta}\{y \leq w\} = \Phi(w) - n^{-1/2}\phi(w)[Q_1(w,\theta) - D_j\{(I^{11})^{-1/2}I^{j1}\}]$$
$$- n^{-1}\phi(w)\Big[Q_2(w,\theta) - \frac{1}{6}(w^3 + 3w)D_v(L_{jrs}\tau^{jr}\tau^{sv})$$
$$+ \frac{1}{2}w\{D_j D_r(\tau^{jr}) - D_v(L_{jrs}\tau^{jr}\sigma^{sv}) - D_r(L_{jsv}\tau^{jr}\sigma^{sv})\}\Big]$$
$$+ o(n^{-1}), \tag{3.2.3}$$

for all θ.

Observe that $P^{\pi}\{y \leq w|X\}$ is stochastic in a frequentist setup. Hence in c.d.f. matching it is appropriate to compare $E_{\theta}[P^{\pi}\{y \leq w|X\}]$ and $P_{\theta}\{y \leq w\}$ for each w and θ. Thus, equating coefficients of $n^{-1/2}$ in the right-hand sides of (3.2.2) and (3.2.3), it follows that a prior $\pi(\cdot)$ ensures first order matching of the posterior and frequentist c.d.f.'s if and only if it satisfies the partial differential equation

$$\Delta_1(\pi, \theta) = 0, \tag{3.2.4}$$

where as in (2.4.8),

$$\Delta_1(\pi, \theta) = D_j\{\pi(\theta)I^{j1}(I^{11})^{-1/2}\}.$$

Similarly, equating coefficients of $n^{-1/2}$, wn^{-1} and w^3n^{-1} in the right-hand sides of (3.2.2) and (3.2.3), a prior $\pi(.)$ ensures matching in the same sense at the second order if and only if it satisfies

$$\Delta_2^{(1)}(\pi, \theta) = 0, \quad \Delta_2^{(2)}(\pi, \theta) = 0, \tag{3.2.5}$$

in addition to (3.2.4), where

$$\Delta_2^{(1)}(\pi, \theta) = D_j D_r\{\tau^{jr}\pi(\theta)\} - 2D_r\{\tau^{jr}\pi_j(\theta)\} - D_v\{L_{jrs}\tau^{jr}\sigma^{sv}\pi(\theta)\}$$
$$- D_r\{L_{jsv}\tau^{jr}\sigma^{sv}\pi(\theta)\}, \tag{3.2.6}$$
$$\Delta_2^{(2)}(\pi, \theta) = D_v\{L_{jrs}\tau^{jr}\tau^{sv}\pi(\theta)\}. \tag{3.2.7}$$

The above results are due to Mukerjee and Ghosh (1997). From (2.4.11) and (3.2.4), it follows that the two approaches, based on quantiles and c.d.f.'s, lead to the same first order matching condition. A comparison of (3.2.5)–(3.2.7) and (2.4.12), however, reveals that the corresponding second order matching conditions are not identical. Analogously to subsection 2.5.3, simpler forms of (3.2.6) and (3.2.7) can be obtained under an orthogonal parmeterization.

From the above discussion, it is clear that the first order matching priors reported in the examples of Chapter 2 (see e.g., Sections 2.5 and 2.6) continue to enjoy the matching property, up to the same order, on the basis of c.d.f.'s. As discussed in Mukerjee and Ghosh (1997), the second order matching conditions (3.2.5) are, however, more restrictive than their counterpart (2.4.12) based on quantiles and often do not have any solution. Hence these second order conditions are not considered any more in what follows.

3.2.2 Single parametric function

Suppose now the interest is in developing c.d.f. matching priors for a one-dimensional smooth function $g(\theta)$ of the parameter vector θ. Since the same invariance argument as in Section 2.7 holds for c.d.f. matching priors as well, in principle, one can reduce $g(\theta)$ to a canonical interest parameter via reparameterization and then apply the results of the previous subsection. As indicated in Section 2.8, such reparameterization may, however, destroy the natural form of the information matrix and occasionally result in matching conditions in the form of awkward partial differential equations. From this point of view, it makes sense to give results directly in terms of $g(\theta)$. Following Datta and Ghosh, J.K. (1995b), the first order c.d.f. matching condition for $g(\theta)$ is presented below. The interested readers are referred to the original article for a proof.

Let $\nabla g(\theta) = (D_1 g(\theta), \ldots, D_p g(\theta))^T$ be the gradient vector of the parametric function $g(\theta)$. Define the vector $\eta \equiv \eta(\theta) = (\eta_1, \ldots, \eta_p)^T$ by

$$\eta = [\{\nabla g(\theta)\}^T I^{-1} \{\nabla g(\theta)\}]^{-1/2} I^{-1} \nabla g(\theta) .$$

Also define $d = [\{\nabla g(\widehat{\theta})\}^T C^{-1} \{\nabla g(\widehat{\theta})\}]^{1/2}$. Then

$$E_\theta P^\pi \left[\frac{n^{1/2}\{g(\theta) - g(\widehat{\theta})\}}{d} \le w | X \right] = P_\theta \left[\frac{n^{1/2}\{g(\theta) - g(\widehat{\theta})\}}{d} \le w \right] + o(n^{-1/2})$$

(3.2.8)

for all w and θ, if and only if $\pi(\cdot)$ satisfies the partial differential equation

$$D_j \{\eta_j \pi(\theta)\} = 0 .$$ (3.2.9)

Note that the quantity $n^{1/2}\{g(\theta) - g(\widehat{\theta})\}/d$, considered in (3.2.8), is an approximate standardized version of $g(\theta)$ in the posterior setup.

It is easy to see that $\eta_j = \Lambda^{-1} b_j$, where b_j and Λ are given by (2.8.1) and (2.8.2) respectively. Hence the last equation is identical with the corresponding partial differential equation (2.8.3) for first order matching based on quantiles. This is well anticipated in view of what was seen in the previous subsection for first order matching with a canonical interest parameter. At this stage, the reader may wonder why (3.2.9) was not expressed at the outset using the same notation as in (2.8.3). The reason for this notational change is that it facilitates the presentation in the next section.

Example 3.2.1. (Datta and Ghosh, J.K., 1995b) Consider the log-normal model given by the density

$$f(x; \theta) = (x\theta_2)^{-1} \phi\left(\frac{\log x - \theta_1}{\theta_2}\right) , \quad x > 0 ,$$

where $\theta_1 \in \mathcal{R}^1$ and $\theta_2 > 0$. Suppose interest lies in $g(\theta) = \exp(\theta_1 + \frac{1}{2}\theta_2^2)$, the population mean. Here

$$I_{11} = \theta_2^{-2} , \quad I_{22} = 2\theta_2^{-2} , \quad I_{12} = 0 .$$

Hence the c.d.f. matching equation (3.2.9) simplifies to

$$D_1\left\{\frac{\theta_2}{(1 + \frac{1}{2}\theta_2^2)^{1/2}}\pi(\theta)\right\} + D_2\left\{\frac{\theta_2^2}{2(1 + \frac{1}{2}\theta_2^2)^{1/2}}\pi(\theta)\right\} = 0 ,$$

with a solution

$$\pi(\theta) = \theta_2^{-2}\left(1 + \frac{1}{2}\theta_2^2\right)^{1/2} .$$

♣

3.3 C.d.f. matching priors for multiple parametric functions

When multiple parameters or parametric functions are of interest, it makes sense to handle them jointly and the c.d.f. matching approach facilitates such a joint treatment. Datta (1996) and Ghosh and Mukerjee (1993a) investigated this problem from two different standpoints. Datta (1996) considered a direct generalization of the setup of subsection 3.2.2 to study c.d.f. matching priors for several parametric functions that are of *equal* interest. On the other hand, Ghosh and Mukerjee (1993a) studied the matching problem for the entire parameter vector, but with an *ordering* of its elements with regard to inferential importance, and based their analysis on some kind of posterior standardized regression residuals. These two approaches are discussed in subsections 3.3.1 and 3.3.2 respectively. The asymptotics in the rest of this chapter are all at the first order of approximation, i.e., with a margin of error $o(n^{-1/2})$, even when this is not mentioned explicitly.

3.3.1 Multiple parametric functions

Let $g_1(\theta), \ldots, g_q(\theta)$ be q ($\leq p$) smooth parametric functions of interest such that the $p \times q$ gradient matrix $(\nabla g_1(\theta), \ldots, \nabla g_q(\theta))$ is of full column rank for all θ. A prior $\pi(\cdot)$ is said to be *joint c.d.f. matching* for $g_1(\theta), \ldots, g_q(\theta)$ if

$$E_\theta P^\pi \left[\frac{n^{1/2}\{g_1(\theta) - g_1(\widehat{\theta})\}}{d_1} \leq w_1, \ldots, \frac{n^{1/2}\{g_q(\theta) - g_q(\widehat{\theta})\}}{d_q} \leq w_q | X \right]$$

$$= P_\theta \left[\frac{n^{1/2}\{g_1(\theta) - g_1(\widehat{\theta})\}}{d_1} \leq w_1, \ldots, \frac{n^{1/2}\{g_q(\theta) - g_q(\widehat{\theta})\}}{d_q} \leq w_q \right]$$

$$+ o(n^{-1/2}) \tag{3.3.1}$$

for all $w_1, \ldots, w_q$ and all θ. In the above, $d_k = [\{\nabla g_k(\widehat{\theta})\}^T C^{-1} \{\nabla g_k(\widehat{\theta})\}]^{1/2}$, $k = 1, \ldots, q$, and $w_1, \ldots, w_q$ are free from n, θ and X. Note that there is no implicit sum in the definition of d_k or in equations (3.3.2) and (3.3.7) below even though a subscript is repeated in the right-hand sides of these equations; this is evident because the same subscript appears also in the respective left-hand sides. The same consideration applies to the subsequent chapters as well.

A prior $\pi(\cdot)$ is said to be *simultaneous marginal c.d.f. matching* for $g_1(\theta), \ldots, g_q(\theta)$ if it is c.d.f. matching separately for each of $g_1(\theta), \ldots, g_q(\theta)$ in the sense of (3.2.8). It follows from (3.2.8) and (3.3.1) that, if $\pi(\cdot)$ is a joint c.d.f. matching prior for $g_1(\theta), \ldots, g_q(\theta)$, then it is also simultaneous marginal c.d.f. matching for $g_1(\theta), \ldots, g_q(\theta)$. This is evident if one keeps any w_k fixed in (3.3.1) and allows $w_m (m \neq k)$ to tend to infinity.

For $1 \leq k, m, u \leq q$, define

$$\epsilon_k \equiv \epsilon_k(\theta) = (\epsilon_{k1}, \ldots, \epsilon_{kp})^T$$
$$= [\{\nabla g_k(\theta)\}^T I^{-1}\{\nabla g_k(\theta)\}]^{-1/2} I^{-1} \nabla g_k(\theta), \qquad (3.3.2)$$

$$\rho_{km} = \epsilon_k^T I \epsilon_m, \qquad (3.3.3)$$

$$\zeta_{kmu} = \epsilon_{kj} D_j \rho_{mu}. \qquad (3.3.4)$$

Also, let R be a $q \times q$ matrix with elements ρ_{km}. Clearly, ϵ_k is analogous to η in subsection 3.2.2 with $g(\theta)$ replaced by $g_k(\theta)$. Note that $\zeta_{kmu} = \zeta_{kum}$. The following theorem, due to Datta (1996), provides the required matching equations.

Theorem 3.3.1. *(a) A prior $\pi(\cdot)$ is simultaneous marginal c.d.f. matching for parametric functions $g_1(\theta), \ldots, g_q(\theta)$ if and only if it satisfies the partial differential equations*

$$D_j\{\epsilon_{kj}\pi(\theta)\} = 0, \quad k = 1, \ldots, q. \qquad (3.3.5)$$

(b) A simultaneous marginal c.d.f. matching prior for parametric functions $g_1(\theta), \ldots, g_q(\theta)$ is joint c.d.f. matching if and only if the conditions

$$\zeta_{kmu} + \zeta_{mku} + \zeta_{ukm} = 0, \quad k, m, u = 1, \ldots, q \qquad (3.3.6)$$

hold.   ♣

Part (a) follows from (3.2.9) by replacing $g(\theta)$ by $g_k(\theta)$, $k = 1, \ldots, q$. The proof of (b) is quite technical, and interested readers are referred to the original source.

Note that the conditions (3.3.6) depend only on the parametric functions and the model. If these conditions hold then simultaneous marginal c.d.f. matching becomes equivalent to joint c.d.f. matching, so that a prior enjoys not only the former but also the latter matching property if and only if it satisfies (3.3.5). The conditions (3.3.6) are trivially satisfied for $q = 1$. Furthermore, it is immediate from (3.3.4) that these are met when the ρ_{km} are constants free from θ, as happens, for instance, in Examples 3.3.1, 3.3.2, 3.3.3 and 3.3.4(b) below. On the other hand, there can also be situations where the conditions (3.3.6) do not hold and, as such, no joint c.d.f. matching prior is available even though simultaneous marginal c.d.f. matching priors may exist. This is illustrated in Example 3.3.4(c) below.

Consider now the special case where interest lies in the entire parameter vector θ, i.e., $q = p$ and $g_k(\theta) = \theta_k$, $1 \le k \le p$, and suppose I is a diagonal matrix for all θ. Then by (3.3.2) and (3.3.3), ρ_{km} equals 1 if $k = m$ and 0 otherwise. Hence the conditions in (3.3.6) hold, so that a prior is joint c.d.f. matching for θ if and only if it is simultaneous marginal c.d.f. matching. Additionally, if

$$I_{jj}(\theta) = e_{1j}(\theta_j)\, e_{2j}(\theta_1, \ldots, \theta_{j-1}, \theta_{j+1}, \ldots, \theta_p) \qquad (3.3.7)$$

for positive functions $e_{1j}(\cdot)$ and $e_{2j}(\cdot)$, $1 \le j \le p$, then by (3.3.5) a unique joint c.d.f. matching prior is given by

$$\pi(\theta) \propto \prod_{j=1}^{p} e_{1j}^{1/2}(\theta_j) \, . \tag{3.3.8}$$

This prior was shown in Datta and Ghosh, M. (1995a) to be a one-at-a-time reference prior for θ.

Example 3.3.1. (Datta, 1996) In the location-scale model (2.5.16), let $q = 2$, $g_1(\theta) = \theta_1$, $g_2(\theta) = \theta_2$ where $\theta_1 \in \mathcal{R}^1$, $\theta_2 > 0$. Then the ρ_{km} are constants and hence the conditions (3.3.6) are satisfied. Therefore, by (3.3.5), the prior $\pi(\theta) \propto \theta_2^{-1}$, considered previously in (2.5.18), is the unique joint c.d.f. matching prior for $g_1(\theta)$ and $g_2(\theta)$. This follows because $I^{jk} \propto \theta_2^2$ for every j, k, under the location-scale model, so that the conditions in (3.3.5) reduce to $D_j\{\theta_2\pi(\theta)\} = 0$ $(j = 1, 2)$; cf. (2.8.10).   ♣

Example 3.3.2. (Datta, 1996) Consider the p-variate normal model with mean vector $\mu = (\mu_1, \ldots, \mu_p)^T$ and dispersion matrix $\mathcal{I}_p$, where $\mathcal{I}_p$ is the identity matrix of order p, and $p \ge 2$. Reparameterize as

$$\mu_1 = \theta_1 \cos\theta_2 \, ,$$
$$\mu_2 = \theta_1 \sin\theta_2 \cos\theta_3 \, ,$$
$$\vdots$$
$$\mu_{p-1} = \theta_1 \sin\theta_2 \ldots \sin\theta_{p-1} \cos\theta_p \, ,$$
$$\mu_p = \theta_1 \sin\theta_2 \ldots \sin\theta_{p-1} \sin\theta_p \, ,$$

where $\theta_1 > 0$, $0 < \theta_2, \ldots, \theta_{p-1} < \pi$, and $0 < \theta_p < 2\pi$, the π appearing in these ranges being the usual transcendental number (not to be confused with a prior). Here

$$I = \text{diag}(1 \, , \, \theta_1^2 \, , \, \theta_1^2 \sin^2\theta_2 \, , \ldots, \, \theta_1^2 \sin^2\theta_2 \ldots \sin^2\theta_{p-1}) \, ,$$

which is in keeping with (3.3.7). Suppose interest lies in the entire parameter vector θ. Then by (3.3.8), $\pi(\theta) = $ constant is the unique joint c.d.f. matching prior.   ♣

Example 3.3.3. (Datta, 1996) We revisit Example 2.6.8 with a somewhat different notation. Consider the balanced one-way random effects model given by

$$X_i^{(j)} = \theta_1 + a_i + e_{ij} \, , \quad 1 \le i \le n \, , \quad 1 \le j \le t \, ,$$

where θ_1 is the general mean, each a_i is univariate normal with mean zero and variance θ_2, each e_{ij} is univariate normal with mean zero and variance θ_3, and the a_i's and the e_{ij}'s are all independent. Here $\theta_1 \in \mathcal{R}^1$ and $\theta_2, \theta_3 > 0$.

Let $q = 3$, $g_1(\theta) = \theta_1$, $g_2(\theta) = \theta_2/\theta_3$ and $g_3(\theta) = \theta_3$. As in Example 2.6.8, writing $X_i = (X_i^{(1)}, \ldots, X_i^{(t)})^T$, it can be checked that

$$I^{11} = \theta_2 + t^{-1}\theta_3 , \quad I^{12} = 0 , \quad I^{13} = 0 ,$$

$$I^{22} = \frac{2\{t(t-1)\theta_2^2 + 2(t-1)\theta_2\theta_3 + \theta_3^2\}}{t(t-1)} , \quad I^{23} = -\frac{2\theta_3^2}{t(t-1)} , \quad I^{33} = \frac{2\theta_3^2}{t-1} ,$$

$$\epsilon_1 = (\theta_2 + t^{-1}\theta_3)^{1/2}(1 , 0 , 0)^T ,$$

$$\epsilon_2 = \left\{\frac{2}{t(t-1)}\right\}^{1/2}(0 , \theta_3 + (t-1)\theta_2 , -\theta_3)^T ,$$

$$\epsilon_3 = \left\{\frac{2}{t^2(t-1)}\right\}^{1/2}\theta_3(0 , -1 , t)^T ,$$

and

$$R = \begin{pmatrix} 1 & 0 & 0 \\ 0 & 1 & -\frac{1}{\sqrt{t}} \\ 0 & -\frac{1}{\sqrt{t}} & 1 \end{pmatrix} ,$$

where, as indicated below (3.3.4), R is the matrix with elements ρ_{km}. Thus the ρ_{km} are constants so that the conditions (3.3.6) are met. From (3.3.5), it now follows that $\pi(\theta) \propto \{\theta_3(\theta_3 + t\theta_2)\}^{-1}$ is the unique joint c.d.f. matching prior for $g_1(\theta)$, $g_2(\theta)$ and $g_3(\theta)$. It was seen earlier in Example 2.6.8 that this is also a second order matching prior for quantiles when the ratio of variance components is the object of interest. ♣

Example 3.3.4. (Datta, 1996) Consider the bivariate normal model of Example 2.8.1 with means θ_1, θ_2, standard deviations θ_3, θ_4 and correlation coefficient θ_5, where $\theta_1, \theta_2 \in \mathcal{R}^1$, $\theta_3, \theta_4 > 0$, and $|\theta_5| < 1$.

(a) First, let $q = 2$ and $g_i(\theta) = \theta_i$, $i = 1, 2$. Using the expressions for the elements of I^{-1} as shown in Example 2.8.1, then

$$\epsilon_1 = (\theta_3 , \theta_4\theta_5 , 0 , 0 , 0)^T , \quad \epsilon_2 = (\theta_3\theta_5 , \theta_4 , 0 , 0 , 0)^T ,$$

and

$$R = \begin{pmatrix} 1 & \theta_5 \\ \theta_5 & 1 \end{pmatrix} .$$

Hence one can check that the conditions (3.3.6) are satisfied. Therefore, by (3.3.5), a prior $\pi(\theta)$ is joint c.d.f. matching if and only if it is free from θ_1 and θ_2.

(b) Next let $q = 3$, $g_1(\theta) = \theta_3$, $g_2(\theta) = \theta_4\theta_5/\theta_3$ and $g_3(\theta) = \theta_4^2(1 - \theta_5^2)$. As before,

$$\epsilon_1 = \frac{1}{\sqrt{2}}(0 , 0 , \theta_3 , \theta_4\theta_5^2 , \theta_5(1 - \theta_5^2))^T ,$$

$$\epsilon_2 = (0 , 0 , 0 , \theta_4\theta_5(1 - \theta_5^2)^{1/2} , (1 - \theta_5^2)^{3/2})^T ,$$

$$\epsilon_3 = \frac{1}{\sqrt{2}}(0 , 0 , 0 , \theta_4(1 - \theta_5^2) , -\theta_5(1 - \theta_5^2))^T$$

and R turns out to be the 3×3 identity matrix. Obviously, the conditions (3.3.6) are satisfied. As in Example 2.8.1, consider a natural class of priors of the form

$$\pi(\theta) = \{\theta_3^{s_3}\theta_4^{s_4}(1 - \theta_5^2)^{s_5}\}^{-1} .$$

From (3.3.5), it can be seen that the unique joint c.d.f. matching prior in this class is $\pi(\theta) = \{\theta_3^2(1 - \theta_5^2)\}^{-1}$.

(c) Finally, let $q = 2$, $g_1(\theta) = \theta_3\theta_5/\theta_4$ and $g_2(\theta) = \theta_4\theta_5/\theta_3$. Then

$$\epsilon_1 = (0 , 0 , \theta_3\theta_5(1 - \theta_5^2)^{1/2} , 0 , (1 - \theta_5^2)^{3/2})^T ,$$

$$\epsilon_2 = (0 , 0 , 0 , \theta_4\theta_5(1 - \theta_5^2)^{1/2} , (1 - \theta_5^2)^{3/2})^T$$

and

$$R = \begin{pmatrix} 1 & 1 - 2\theta_5^2 \\ 1 - 2\theta_5^2 & 1 \end{pmatrix} .$$

Hence it is not hard to see that conditions (3.3.6) fail. Thus there is no joint c.d.f. matching prior in this case. However, (3.3.5) shows that any prior of the form $\pi(\theta) \propto (\theta_3\theta_4)^{-r}(1-\theta_5^2)^{-(1+\frac{1}{2}r)}$, where r is a real number, is simultaneous marginal c.d.f. matching. The choice $r = 1$ results in the prior $\pi(\theta) \propto \{\theta_3\theta_4(1 - \theta_5^2)^{3/2}\}^{-1}$, which has received some attention in other contexts (Geisser, 1965; Datta and Ghosh, J.K., 1995b).   ♣

Example 3.3.5. (Ghosh and Mukerjee, 1993a; Datta, 1996) This example, though somewhat contrived, demonstrates that even under orthogonal parameterization, no simultaneous marginal c.d.f. matching prior may exist and hence no joint c.d.f. matching prior may be available. Let

$$f(x;\theta) = \theta_2^{-1/2}\phi\left(\theta_2^{-1/2}(x^{(1)} - \theta_1)\right) \exp\left\{ -(\theta_1^3 + x^{(2)}e^{-\theta_1^3})\right\} ,$$

where $x = (x^{(1)}, x^{(2)})^T$, $x^{(1)} \in \mathcal{R}^1$, $x^{(2)} > 0$, and $\theta_1 \in \mathcal{R}^1$, $\theta_2 > 0$. Here $I = \mathrm{diag}(\theta_2^{-1} + 9\theta_1^4, \frac{1}{2}\theta_2^{-2})$. Let $q = 2$, $g_1(\theta) = \theta_1$, $g_2(\theta) = \theta_2$. Upon checking (3.3.5), then it easily follows that no simultaneous marginal c.d.f. matching prior exists.   ♣

3.3.2 Regression residuals approach to c.d.f. matching

Ghosh and Mukerjee (1993a) considered matching priors for the entire parameter vector $\theta = (\theta_1, \ldots, \theta_p)^T$ when the θ_i's are of decreasing importance in i. This kind of ordering of the θ_i's is common in the context of reference priors and may be contrasted with the setup of the last subsection where the parametric functions were of equal interest.

The observed information matrix $C = ((c_{jr}))$, introduced in Section 2.2, plays a key role in the approach of Ghosh and Mukerjee (1993a). Recall that C is positive definite over the set on which posterior expansions are being considered. Hence there exists a unique lower triangular matrix C^*, with diagonal elements all positive, such that

$$C = C^{*T}C^* . \tag{3.3.9}$$

For example, if $p = 2$ then

$$C^* = \begin{bmatrix} (c_{11} - c_{22}^{-1}c_{12}^2)^{1/2} & 0 \\ c_{22}^{-1/2}c_{12} & c_{22}^{1/2} \end{bmatrix} .$$

Ghosh and Mukerjee (1993a) based their analysis on the pivotal quantity

$$W = (W_1 ,\dots , W_p)^T = n^{1/2}C^*(\theta - \widehat{\theta}) , \tag{3.3.10}$$

and matched the posterior and frequentist c.d.f.'s of W. This can be motivated as follows. By (2.2.19), $h = n^{1/2}(\theta - \widehat{\theta})$ is asymptotically p−variate normal in the posterior setup, with null mean vector and dispersion matrix C^{-1}. Hence by (3.3.9) and (3.3.10), it is possible to interpret W_1 as an approximate posterior standardized version of θ_1. In fact, it is easily seen that W_1 is the same as $y = (n/c^{11})^{1/2}(\theta_1 - \widehat{\theta}_1)$ considered in subsection 3.2.1. Similarly, for $2 \le i \le p$, one can interpret W_i as an approximate standardized version of the regression residual of θ_i on $\theta_1,\dots,\theta_{i-1}$ in the posterior setup. Ghosh and Mukerjee (1993a) argued that consideration of such residuals is natural when the θ_i's are of decreasing importance in i.

In the above approach based on regression residuals, a prior $\pi(\cdot)$ is said to be c.d.f. matching if

$$E_\theta[P^\pi\{W_1 \le w_1 ,\dots , W_p \le w_p|X\}]$$
$$= P_\theta\{W_1 \le w_1 ,\dots , W_p \le w_p\} + o(n^{-1/2}) \tag{3.3.11}$$

for all $w_1,\dots,w_p$ and all θ. As before, here $w_1,\dots,w_p$ are free from n,θ and X. Ghosh and Mukerjee (1993a) gave the following characterization for c.d.f. matching priors as envisaged in (3.3.11). We refer to the original source for a proof.

Theorem 3.3.2. *A prior $\pi(\cdot)$ is c.d.f. matching in the approach based on regression residuals if and only if it satisfies the partial differential equations*

$$D_j\{\widetilde{I}_{jk}\pi(\theta)\} = 0 , \quad k = 1,\dots,p , \tag{3.3.12}$$

where $\widetilde{I} = ((\widetilde{I}_{jk})) = (I^*)^{-1}$, and I^* is defined with reference to I, exactly as C^* was defined with reference to C. ♣

Example 3.3.6. (Ghosh and Mukerjee, 1993a) For the multivariate location model

$$f(x; \theta) = f^*(x^{(1)} - \theta_1, \ldots, x^{(p)} - \theta_p),$$

where $x = (x^{(1)}, \ldots, x^{(p)})^T$ and $\theta = (\theta_1, \ldots, \theta_p)^T \in \mathcal{R}^p$, it is easily seen that the I_{jk} and hence the $\widetilde{I}_{jk}$ are all constants. Therefore, the unique prior satisfying (3.3.12) is given by $\pi(\theta) = $ constant, which is also Jeffreys' prior.

Similarly, for the multivariate scale model

$$f(x; \theta) = (\theta_1 \ldots \theta_p)^{-1} f^*(x^{(1)}/\theta_1, \ldots, x^{(p)}/\theta_p),$$

where $x = (x^{(1)}, \ldots, x^{(p)})^T$ and $\theta_j > 0, j = 1, \ldots, p$, one gets $I_{jk} \propto (\theta_j \theta_k)^{-1}$ for every j, k. Hence $\widetilde{I}_{jk} \propto \theta_j$ for every j, k, and $\pi(\theta) \propto (\theta_1 \ldots \theta_p)^{-1}$ emerges as the unique solution to (3.3.12). Note that this is again Jeffreys' prior. ♣

Example 3.3.7. (Ghosh and Mukerjee, 1993a) For the location-scale model considered previously in Example 3.3.1, $I_{jk} \propto \theta_2^{-2}$, and hence $\widetilde{I}_{jk} \propto \theta_2$ for every j, k. Consequently, $\pi(\theta) \propto \theta_2^{-1}$ is the unique solution to (3.3.12). Recall that the same prior was obtained in (2.5.18) and Example 3.3.1 from other considerations. Note that this is not Jeffreys' prior. ♣

In particular, if I is a diagonal matrix for all θ, then the matching conditions (3.3.12) reduce to

$$D_1\{I_{11}^{-1/2}\pi(\theta)\} = 0, \ldots, D_p\{I_{pp}^{-1/2}\pi(\theta)\} = 0. \tag{3.3.13}$$

From (3.3.2), it is readily seen that in this situation the matching conditions (3.3.5) also reduce to (3.3.13) when one takes $g_k(\theta) = \theta_k, 1 \le k \le p$. Furthermore, as noted in the previous subsection, the conditions (3.3.6) also hold. As a result, if I is a diagonal matrix and interest lies in the entire parameter vector θ, then the three concepts of (a) simultaneous marginal c.d.f. matching, (b) joint c.d.f. matching, and (c) c.d.f. matching via regression residuals become equivalent. It is now immediate that no c.d.f. matching prior on the basis of regression residuals exists in Example 3.3.5. For further discussion connecting these three concepts, the reader may see Datta (1996).

Example 3.3.8. (Ghosh and Mukerjee, 1993a) We revisit the exponential regression model of Example 2.6.5. Then I is a diagonal matrix, I_{11} is a constant and $I_{22} \propto \theta_2^{-2}$. Thus $\pi(\theta) \propto \theta_2^{-1}$ is the unique solution to (3.3.13) and hence to (3.3.12). The same prior was obtained in Example 2.6.5 from consideration of posterior quantiles. ♣

One can check that the c.d.f. matching priors reported in Examples 3.2.1, 3.3.1–3.3.4 and 3.3.6–3.3.8 entail the propriety of the posteriors, with P_θ–probability unity for all θ, whenever n is sufficiently large. For some of these priors, considered earlier in Chapter 2, this has already been indicated in Section 2.10. Incidentally, no c.d.f. matching prior was available in Example 3.3.5.

Matching Priors for Highest Posterior Density Regions

4.1 Introduction

Highest posterior density (HPD) regions are very popular with Bayesians. With a possibly multidimensional interest parameter $\widetilde{\theta}$, such a region is of the form

$$\{\widetilde{\theta} : \pi(\widetilde{\theta}|X) \geq K\} ,$$

where $\pi(\widetilde{\theta}|X)$ is the posterior density of $\widetilde{\theta}$, under a prior $\pi(\cdot)$, given the data X, and K depends on $\pi(\cdot)$ and X in addition to the chosen posterior credibility level. Clearly, by the Neyman-Pearson lemma, an HPD region has the smallest possible volume, given X, at a chosen level of credibility. In this chapter, we consider priors ensuring approximate frequentist validity of HPD regions with margin of error $o(n^{-1})$, where n is the sample size. Priors of this kind are called matching priors for HPD regions or, briefly, HPD matching priors. They can be useful even when the interest parameter is multidimensional since HPD regions are well-defined in such situations.

The next two sections aim at studying HPD matching priors in the absence of nuisance parameters; the interest parameter is allowed to be possibly multidimensional. In Section 4.2, an explicit form of an HPD region is worked out and the matching conditions are derived in Section 4.3, where the issue of invariance is also addressed. Further results in the presence of nuisance parameters are indicated and discussed in Section 4.4.

4.2 Explicit form of an HPD region

We continue with the setup and assumptions of Section 2.2. Throughout this chapter, unless otherwise specified, the notation is also as described there. Suppose interest lies in the entire parameter vector $\theta = (\theta_1, \ldots, \theta_p)^T$, i.e., there is no nuisance parameter. The posterior density of θ, under a prior $\pi(\cdot)$, given $X = (X_1, \ldots, X_n)^T$ satisfies

$$\pi(\theta|X) \propto \pi(\theta)\exp[n\{\ell(\theta) - \ell(\widehat{\theta})\}] \, ,$$

where the constant of proportionality depends on $\pi(\cdot)$ and X but not on θ. Hence considering $-2\log\pi(\theta|X)$, an HPD region for θ is given by

$$Q^{(1-\alpha)}(\pi, X) = \{\theta : M(\theta, \pi, X) \le k_{1-\alpha}(\pi, X)\} \, , \qquad (4.2.1)$$

where

$$M(\theta, \pi, X) = -2\{\log\pi(\theta) - \log\widehat{\pi}\} - 2n\{\ell(\theta) - \ell(\widehat{\theta})\} + n^{-1}W_0 \, , \quad (4.2.2)$$

with

$$W_0 = (\widehat{\pi}_j\widehat{\pi}_r/\widehat{\pi}^2)c^{jr} \, , \qquad (4.2.3)$$

and $k_{1-\alpha}(\pi, X)$, which may depend on $\pi(\cdot)$ and X but not on θ, has to be so chosen that the relation

$$P^{\pi}\{\theta \in Q^{(1-\alpha)}(\pi, X)|X\} = 1 - \alpha + o(n^{-1}) \qquad (4.2.4)$$

holds. As usual, here $0 < \alpha < 1$ and $P^{\pi}\{\cdot|X\}$ is the posterior probability measure under the prior $\pi(\cdot)$. Since neither $\log\widehat{\pi}$ nor W_0 involves θ, their inclusion in (4.2.2) and hence in (4.2.1) does not affect the HPD status of $Q^{(1-\alpha)}(\pi, X)$. Rather, as seen below, this helps in finding an expression for $k_{1-\alpha}(\pi, X)$ in a neat form.

An explicit determination of the HPD region considered in (4.2.1) warrants that of $k_{1-\alpha}(\pi, X)$. This, in turn, necessitates studying the posterior distribution of $M(\theta, \pi, X)$ up to the order of approximation $o(n^{-1})$. To that effect, we have Lemma 4.2.1 below. In what follows, $E^{\pi}[\cdot|X]$ denotes posterior expectation under the prior $\pi(\cdot)$. Also, we write $\xi = (-1)^{1/2}t$ where t is an auxiliary variable, and define $W_1, \ldots, W_4$ via (2.2.18).

Lemma 4.2.1. *Under the prior $\pi(\cdot)$, the approximate posterior characteristic function of $M(\theta, \pi, X)$ is given by*

$$E^{\pi}[\exp\{\xi M(\theta, \pi, X)\}|X]$$

$$= (1 - 2\xi)^{-\frac{1}{2}p}\left[1 + \frac{1}{24}n^{-1}\left(\frac{1}{1 - 2\xi} - 1\right)\left(W_3 + \frac{1}{3}W_4\right)\right] + o(n^{-1}) \, .$$

Proof. As in Section 2.2, let $h = (h_1, \ldots, h_p)^T = n^{1/2}(\theta - \widehat{\theta})$. Then by (2.2.10), (2.2.11) and (4.2.2),

$$M(\theta, \pi, X) = h^T Ch - n^{-1/2}\left\{2R_1(h) + \frac{1}{3}R_3(h)\right\}$$

$$+ n^{-1}\left[\{R_1(h)\}^2 - R_2(h) - \frac{1}{12}R_4(h) + W_0\right]$$

$$+ o(n^{-1}) \, , \qquad (4.2.5)$$

where $R_1(h), \ldots, R_4(h)$ are as given by (2.2.12). Hence, using (2.2.19), after some algebra

$$
E^\pi[\exp\{\xi M(\theta, \pi, X)\}|X]
$$
$$
= \int \exp\{\xi M(\theta, \pi, X)\}\pi^*(h|X)dh
$$
$$
= \int \left[1 + n^{-1/2}(1-2\xi)\left\{R_1(h) + \frac{1}{6}R_3(h)\right\}\right.
$$
$$
+ n^{-1}\left\{\xi W_0 - \xi(1-2\xi)\left(R_1(h)\right)^2 + \frac{1}{2}(1-2\xi)R_2(h)\right.
$$
$$
+ \frac{1}{72}(1-2\xi)^2\left(R_3(h)\right)^2 + \frac{1}{6}(1-2\xi)^2 R_1(h)R_3(h) + \frac{1}{24}(1-2\xi)R_4(h)
$$
$$
\left.\left. - \frac{1}{2}W_1 - \frac{1}{6}W_2 - \frac{1}{24}W_3 - \frac{1}{72}W_4\right\}\right]e^{\xi h^T Ch}\phi_p(h; C^{-1})dh
$$
$$
+ o(n^{-1}) . \tag{4.2.6}
$$

Now,

$$
\int e^{\xi h^T Ch}\phi_p(h; C^{-1})dh = (1-2\xi)^{-\frac{1}{2}p} ,
$$

while by (2.2.12), (2.2.15), (2.2.18) and (4.2.3),

$$
\int \left\{R_1(h) + \frac{1}{6}R_3(h)\right\}e^{\xi h^T Ch}\phi_p(h; C^{-1})dh = 0 ,
$$

$$
\int \{R_1(h)\}^2 e^{\xi h^T Ch}\phi_p(h; C^{-1})dh = (1-2\xi)^{-(\frac{1}{2}p+1)}W_0 ,
$$

$$
\int R_2(h)e^{\xi h^T Ch}\phi_p(h; C^{-1})dh = (1-2\xi)^{-(\frac{1}{2}p+1)}W_1 ,
$$

$$
\int \{R_3(h)\}^2 e^{\xi h^T Ch}\phi_p(h; C^{-1})dh = (1-2\xi)^{-(\frac{1}{2}p+3)}W_4 ,
$$

$$
\int R_1(h)R_3(h)e^{\xi h^T Ch}\phi_p(h; C^{-1})dh = (1-2\xi)^{-(\frac{1}{2}p+2)}W_2 ,
$$

$$
\int R_4(h)e^{\xi h^T Ch}\phi_p(h; C^{-1})dh = (1-2\xi)^{-(\frac{1}{2}p+2)}W_3 .
$$

Using the above in (4.2.6), the lemma follows.   ♣

Let $\Psi_j(\cdot)$ and $\psi_j(\cdot)$ denote respectively the cumulative distribution function and density function of a chi-square variate with j degrees of freedom $(j = 1, 2, \ldots)$. Also, let z^2 be the $(1-\alpha)$th quantile of the chi-square distribution with p degrees of freedom. We shall show that if

$$
k_{1-\alpha}(\pi, X) = z^2\left\{1 + \frac{1}{12}(np)^{-1}\left(W_3 + \frac{1}{3}W_4\right)\right\}, \tag{4.2.7}
$$

then (4.2.4) holds. This will be proved by inverting the approximate posterior characteristic function given in Lemma 4.2.1, a step that can be justified following Chandra and Ghosh (1979) or Chandra (1980). Under (4.2.7), thus one gets

$$P^\pi\{M(\theta,\pi,X) \le k_{1-\alpha}(\pi,X)|X\}$$
$$= \Psi_p(k_{1-\alpha}(\pi,X)) + \frac{1}{24}n^{-1}\left(W_3 + \frac{1}{3}W_4\right)$$
$$\times \{\Psi_{p+2}(k_{1-\alpha}(\pi,X)) - \Psi_p(k_{1-\alpha}(\pi,X))\} + o(n^{-1}), \qquad (4.2.8)$$

and

$$\Psi_p(k_{1-\alpha}(\pi,X)) = \Psi_p(z^2) + \frac{1}{12}(np)^{-1}z^2\left(W_3 + \frac{1}{3}W_4\right)\psi_p(z^2) + o(n^{-1}), \quad (4.2.9)$$

$$\Psi_{p+2}(k_{1-\alpha}(\pi,X)) - \Psi_p(k_{1-\alpha}(\pi,X)) = \Psi_{p+2}(z^2) - \Psi_p(z^2) + o(1)$$
$$= -2p^{-1}z^2\psi_p(z^2) + o(1). \quad (4.2.10)$$

Since $\Psi_p(z^2) = 1 - \alpha$, from (4.2.9) and (4.2.10), it is evident that the right-hand side of (4.2.8) equals $1 - \alpha + o(n^{-1})$. Together with (4.2.1), this implies that if $k_{1-\alpha}(\pi,X)$ is as given by (4.2.7) then indeed (4.2.4) holds. Thus the HPD region $Q^{(1-\alpha)}(\pi,X)$ is explicitly described by (4.2.1) and (4.2.7).

Incidentally, the structure of the approximate posterior characteristic function, as described in Lemma 4.2.1, also implies posterior Bartlett adjustability of $M(\theta,\pi,X)$. This, however, will not be required in the sequel and we refer to DiCiccio and Stern (1993) for details.

4.3 Characterization of HPD matching priors

We now calculate the frequentist coverage probability $P_\theta\{\theta \in Q^{(1-\alpha)}(\pi,X)\}$, up to the order of approximation $o(n^{-1})$, with the objective of characterizing HPD matching priors for θ. Steps 1–3 of Section 1.2 again facilitate the computation. These are implemented as in Sections 1.3 and 2.4.

Step 1: We first consider $P^{\overline{\pi}}\{\theta \in Q^{(1-\alpha)}(\pi,X)|X\}$, where the auxiliary prior $\overline{\pi}(\cdot)$, satisfying the conditions in Bickel and Ghosh (1990), is as described in Section 2.4. It again helps to obtain the approximate posterior characteristic function of $M(\theta,\pi,X)$ under $\overline{\pi}(\cdot)$. By (4.2.5) and a counterpart of (2.2.19), with the implicit $\pi(\cdot)$ there replaced by $\overline{\pi}(\cdot)$, one gets

$$E^{\overline{\pi}}[\exp\{\xi M(\theta,\pi,X)\}|X]$$
$$= (1-2\xi)^{-\frac{1}{2}p}\left[1 + n^{-1}\left(\frac{1}{1-2\xi} - 1\right)\left\{\frac{1}{24}\left(W_3 + \frac{1}{3}W_4\right)\right.\right.$$
$$+ \frac{1}{6}(\overline{W}_2 - W_2) + \frac{1}{2}(\overline{W}_1 - W_1) + (W_0 - W_0^*)\Big\}\Big]$$
$$+ o(n^{-1}), \qquad (4.3.1)$$

where

$$W_0^* = \hat{\pi}_j \widehat{\overline{\pi}}_r c^{jr}/(\hat{\pi}\widehat{\overline{\pi}}) , \qquad (4.3.2)$$

and, analogously to (2.2.18), $\overline{W}_1$ and $\overline{W}_2$ are defined with reference to the prior $\overline{\pi}(\cdot)$ as

$$\overline{W}_1 = \widehat{\overline{\pi}}_{jr} c^{jr}/\widehat{\overline{\pi}} , \quad \overline{W}_2 = 3a_{jrs}\widehat{\overline{\pi}}_u c^{jr} c^{su}/\widehat{\overline{\pi}} . \qquad (4.3.3)$$

The derivation of (4.3.1) is quite is similar to that of Lemma 4.2.1. By (4.2.1), (4.2.9) and (4.2.10), inversion of (4.3.1) yields

$$P^{\overline{\pi}}\{\theta \in Q^{(1-\alpha)}(\pi, X)|X\}$$

$$= \Psi_p(k_{1-\alpha}(\pi, X)) + n^{-1}\{\frac{1}{24}\left(W_3 + \frac{1}{3}W_4\right) + \frac{1}{6}(\overline{W}_2 - W_2) + \frac{1}{2}(\overline{W}_1 - W_1)$$

$$+ (W_0 - W_0^*)\}\{\Psi_{p+2}(k_{1-\alpha}(\pi, X)) - \Psi_p(k_{1-\alpha}(\pi, X))\} + o(n^{-1})$$

$$= 1 - \alpha + 2(np)^{-1} z^2 \psi_p(z^2)\{\frac{1}{6}(W_2 - \overline{W}_2) + \frac{1}{2}(W_1 - \overline{W}_1) + (W_0^* - W_0)\}$$

$$+ o(n^{-1}) . \qquad (4.3.4)$$

Step 2: Recalling (2.2.1)–(2.2.3) and (2.2.6), from (2.2.18), (4.2.3) and (4.3.2)–(4.3.4) one gets

$$E_\theta[P^{\overline{\pi}}\{\theta \in Q^{(1-\alpha)}(\pi, X)|X\}]$$

$$= 1 - \alpha + 2(np)^{-1} z^2 \psi_p(z^2)\{\frac{1}{2}L_{jrs}I^{jr}I^{su}\left(\frac{\pi_u(\theta)}{\pi(\theta)} - \frac{\overline{\pi}_u(\theta)}{\overline{\pi}(\theta)}\right)$$

$$+ \frac{1}{2}I^{jr}\left(\frac{\pi_{jr}(\theta)}{\pi(\theta)} - \frac{\overline{\pi}_{jr}(\theta)}{\overline{\pi}(\theta)}\right) + I^{jr}\frac{\pi_j(\theta)}{\pi(\theta)}\left(\frac{\overline{\pi}_r(\theta)}{\overline{\pi}(\theta)} - \frac{\pi_r(\theta)}{\pi(\theta)}\right)\}$$

$$+ o(n^{-1}) , \qquad (4.3.5)$$

for θ in the interior of the support of $\overline{\pi}(\cdot)$.

Step 3: We proceed exactly as in the corresponding step in Section 2.4. By (4.3.5), this yields

$$P_\theta\{\theta \in Q^{(1-\alpha)}(\pi, X)\} = 1 - \alpha + 2(np)^{-1}\frac{z^2\psi_p(z^2)}{\pi(\theta)}\Delta(\pi, \theta) + o(n^{-1}) , \quad (4.3.6)$$

for all θ, where

$$\Delta(\pi, \theta) = \frac{1}{2}L_{jrs}I^{jr}I^{su}\pi_u(\theta) + \frac{1}{2}\pi(\theta)D_u(L_{jrs}I^{jr}I^{su})$$

$$+ \frac{1}{2}I^{jr}\pi_{jr}(\theta) - \frac{1}{2}\pi(\theta)D_jD_r(I^{jr})$$

$$- \pi(\theta)D_r\{I^{jr}\pi_j(\theta)/\pi(\theta)\} - I^{jr}\pi_j(\theta)\pi_r(\theta)/\pi(\theta)$$

$$= \frac{1}{2}D_u\{\pi(\theta)L_{jrs}I^{jr}I^{su}\}$$

$$+ \frac{1}{2}I^{jr}\pi_{jr}(\theta) - \frac{1}{2}\pi(\theta)\mathrm{D}_j\mathrm{D}_r(I^{jr}) - \mathrm{D}_r\{I^{jr}\pi_j(\theta)\}$$

$$= \frac{1}{2}\mathrm{D}_u\{\pi(\theta)L_{jrs}I^{jr}I^{su}\} - \frac{1}{2}\mathrm{D}_j\mathrm{D}_r\{\pi(\theta)I^{jr}\} . \tag{4.3.7}$$

The last step in (4.3.7) follows as

$$\frac{1}{2}I^{jr}\pi_{jr}(\theta) - \frac{1}{2}\pi(\theta)\mathrm{D}_j\mathrm{D}_r(I^{jr}) - \mathrm{D}_r\{I^{jr}\pi_j(\theta)\}$$

$$= \frac{1}{2}I^{jr}\pi_{jr}(\theta) - \frac{1}{2}\pi(\theta)\mathrm{D}_j\mathrm{D}_r(I^{jr}) - \pi_j(\theta)\mathrm{D}_r(I^{jr}) - I^{jr}\pi_{jr}(\theta)$$

$$= -\frac{1}{2}\{\pi(\theta)\mathrm{D}_j\mathrm{D}_r(I^{jr}) + 2\pi_j(\theta)\mathrm{D}_r(I^{jr}) + I^{jr}\pi_{jr}(\theta)\}$$

$$= -\frac{1}{2}\{\pi(\theta)\mathrm{D}_j\mathrm{D}_r(I^{jr}) + \pi_j(\theta)\mathrm{D}_r(I^{jr}) + \pi_r(\theta)\mathrm{D}_j(I^{jr}) + I^{jr}\pi_{jr}(\theta)\}$$

$$= -\frac{1}{2}\mathrm{D}_j\mathrm{D}_r\{\pi(\theta)I^{jr}\} , \tag{4.3.8}$$

using the invariance of I^{jr} under permutation of its superscripts.

We are now in a position to give a characterization for HPD matching priors when interest lies in the entire parameter vector θ. As hinted in Section 4.1, these priors are required to ensure

$$P_\theta\{\theta \in Q^{(1-\alpha)}(\pi, X)\} = 1 - \alpha + o(n^{-1})$$

for all α and θ. From (4.3.6) and (4.3.7), the following result is evident.

Theorem 4.3.1. *A prior $\pi(\cdot)$ is HPD matching for θ if and only if it satisfies the partial differential equation*

$$D_u\{\pi(\theta)L_{jrs}I^{jr}I^{su}\} - D_jD_r\{\pi(\theta)I^{jr}\} = 0 . \tag{4.3.9}$$

♣

The above result is due to Ghosh and Mukerjee (1993b) who reported it in another equivalent form. Earlier, Peers (1968) and Severini (1991) explored HPD matching priors for scalar θ. Then $p = 1$, $\theta = \theta_1$, I becomes a scalar, and (4.3.9) becomes

$$\frac{\mathrm{d}}{\mathrm{d}\theta}\{\pi(\theta)L_{111}I^{-2}\} - \frac{\mathrm{d}^2}{\mathrm{d}\theta^2}\{\pi(\theta)I^{-1}\} = 0 ,$$

$$\text{i.e.,} \quad \pi(\theta)L_{111}I^{-2} - \frac{\mathrm{d}}{\mathrm{d}\theta}\{\pi(\theta)I^{-1}\} = \text{constant} .$$

Using the regularity condition (2.5.4), it is easily seen that the above is equivalent to

$$I^{-1}\left(\frac{\mathrm{d}\pi(\theta)}{\mathrm{d}\theta}\right) + \pi(\theta)L_{1,11}I^{-2} = \text{constant} . \tag{4.3.10}$$

The HPD matching condition (4.3.10), arising for $p = 1$, was reported in Peers (1968). One can check that it is also equivalent to the corresponding condition given in Severini (1991).

Continuing with $p = 1$ and using (2.5.4) again, a prior of the form $\pi(\theta) \propto I^r$, where r is a real number, satisfies (4.3.10) if and only if

$$I^{r-2}\{(1-r)L_{1,11} - rL_{111}\} = \text{constant} . \tag{4.3.11}$$

In particular, taking $r = 1/2$ in the above, (4.3.10) holds for Jeffreys' prior if and only if

$$I^{-3/2}(L_{1,11} - L_{111}) = \text{constant} . \tag{4.3.12}$$

The condition (4.3.12) holds for the one-parameter location or scale models introduced in (2.5.6) and (2.5.11) respectively. This follows noting that each of I, $L_{1,11}$ and L_{111} is a constant, free from θ, for the location model, and that $I \propto \theta^{-2}$, $L_{1,11} \propto \theta^{-3}$, $L_{111} \propto \theta^{-3}$ for the scale model. Thus, for these models, Jeffreys' prior is HPD matching for θ.

We now present several examples illustrating HPD matching priors. The cases of both scalar and vector θ are considered.

Example 4.3.1. (Ghosh and Mukerjee, 1993b) This example illustrates that even beyond the standard location or scale models Jeffreys' prior can enjoy the HPD matching property. In the spirit of Example 2.5.1, consider the bivariate normal regression model

$$f(x;\theta) = \phi(x^{(1)})\phi(x^{(2)} - \theta x^{(1)}) , \quad x = (x^{(1)}, x^{(2)})^T \in \mathcal{R}^2 ,$$

where $\theta \in \mathcal{R}^1$. Then $I = 1$, $L_{1,11} = L_{111} = 0$, so that (4.3.12) holds and Jeffreys' prior is HPD matching for θ. ♣

Example 4.3.2. We now demonstrate that, even with $p = 1$, Jeffreys' prior does not always enjoy the HPD matching property. Consider the bivariate normal model with zero means, unit variances and correlation coefficient θ, where $|\theta| < 1$. Then

$$I = \frac{1+\theta^2}{(1-\theta^2)^2} , \quad L_{1,11} = -\frac{1}{2}L_{111} = \frac{2\theta(3+\theta^2)}{(1-\theta^2)^3} .$$

Then (4.3.12) does not hold but (4.3.11) is satisfied by $r = -1$. Hence Jeffreys' prior is not HPD matching for θ but $\pi(\theta) \propto I^{-1}$ enjoys this property. This may be contrasted with the findings in Examples 2.5.2 where, with the same model, it was seen that no second order matching prior for the posterior quantiles of θ exists. ♣

Example 4.3.3. (Ghosh and Mukerjee, 1993b) Consider the location-scale model introduced in (2.5.16). Then $p = 2$ and the I_{jr} and the L_{jrs} are as in (2.5.17). Consequently, $\pi(\theta) \propto \theta_2^{-1}$ satisfies (4.3.9) and is HPD matching for $\theta = (\theta_1, \theta_2)^T$. Recall that the same prior was obtained in subsection 2.5.2 from consideration of posterior quantiles and in Examples 3.3.1 and 3.3.7 from consideration of c.d.f. ♣

Example 4.3.4. We revisit the exponential regression model of Example 2.6.5. Then

$$I_{11} = \sum_{j=1}^{t} z_j^2 , \quad I_{22} = t/\theta_2^2 , \quad I_{12} = 0 ,$$

$$L_{111} = \sum_{j=1}^{t} z_j^3 , \quad L_{112} = \sum_{j=1}^{t} z_j^2/\theta_2 , \quad L_{122} = 0 , \quad L_{222} = 4t/\theta_2^3 .$$

Hence the same prior as in Examples 2.6.5 and 3.3.8, namely $\pi(\theta) \propto \theta_2^{-1}$, satisfies (4.3.9) and is HPD matching for $\theta = (\theta_1, \theta_2)^T$.   ♣

Example 4.3.5. Consider the inverse Gaussian model

$$f(x; \theta) = \{\theta_2/(2\pi x^3)\}^{1/2} \exp \left\{ -\frac{1}{2}\theta_2 (x - \theta_1)^2/(\theta_1^2 x) \right\} , \quad x > 0 ,$$

in a natural parameterization, where θ_1, $\theta_2 > 0$ and π is the usual transcendental number (not to be confused with a prior). Then

$$I_{11} = \theta_1^{-3}\theta_2 , \quad I_{22} = \frac{1}{2}\theta_2^{-2} , \quad I_{12} = 0 ,$$

$$L_{111} = 6\theta_1^{-4}\theta_2 , \quad L_{112} = -\theta_1^{-3} , \quad L_{122} = 0 , \quad L_{222} = \theta_2^{-3} .$$

Hence the prior $\pi(\theta) \propto (\theta_1^2\theta_2)^{-1}$ satisfies (4.3.9) and is HPD matching for $\theta = (\theta_1, \theta_2)^T$.   ♣

Example 4.3.6. Consider the gamma model

$$f(x; \theta) = x^{\theta_1 - 1} e^{-x/\theta_2}/\{\theta_2^{\theta_1} \Gamma(\theta_1)\} , \quad x > 0 ,$$

in a natural parameterization, where θ_1, $\theta_2 > 0$. Then

$$I_{11} = \frac{d^2}{d\theta_1^2} \log \Gamma(\theta_1) , \quad I_{22} = \theta_1 \theta_2^{-2} , \quad I_{12} = \theta_2^{-1} ,$$

$$L_{111} = -\frac{d^3}{d\theta_1^3} \log \Gamma(\theta_1) , \quad L_{112} = 0 , \quad L_{122} = \theta_2^{-2} , \quad L_{222} = 4\theta_1 \theta_2^{-3} .$$

Hence the prior $\pi(\theta) \propto (\theta_1^3\theta_2)^{-1}$ satisfies (4.3.9) and is HPD matching for $\theta = (\theta_1, \theta_2)^T$.   ♣

Example 4.3.7. Suppose interest lies simultaneously in the ratio and the product of the means of two independent exponential distributions. Then we consider

$$f(x; \theta) = (\mu_1 \mu_2)^{-1} \exp \left[-\left\{ \frac{x^{(1)}}{\mu_1} + \frac{x^{(2)}}{\mu_2} \right\} \right] , \quad x^{(1)} , x^{(2)} > 0 , \quad (4.3.13)$$

where $x = (x^{(1)}, x^{(2)})^T$, $\mu_1 \equiv \mu_1(\theta) = (\theta_1\theta_2)^{1/2}$, $\mu_2 \equiv \mu_2(\theta) = (\theta_2/\theta_1)^{1/2}$, and θ_1, $\theta_2 > 0$. Note that $\theta_1 = \mu_1/\mu_2$ and $\theta_2 = \mu_1\mu_2$ are the objects of interest. Under the $\theta-$parameterization,

$$I_{11} = \frac{1}{2}\theta_1^{-2}, \quad I_{22} = \frac{1}{2}\theta_2^{-2}, \quad I_{12} = 0,$$

$$L_{111} = \frac{3}{2}\theta_1^{-3}, \quad L_{112} = \frac{1}{4}\theta_1^{-2}\theta_2^{-1}, \quad L_{122} = 0, \quad L_{222} = \frac{7}{4}\theta_2^{-3},$$

Hence the prior $\pi(\theta) \propto (\theta_1\theta_2)^{-1}$ satisfies (4.3.9) and is HPD matching for $\theta = (\theta_1, \theta_2)^T$. ♣

One can check that the HPD matching priors obtained in the above examples entail the propriety of the posteriors, with $P_\theta-$probability unity for all θ, whenever n is sufficiently large. In quite a few of these examples, other priors satisfying the HPD matching condition (4.3.9) exist. Thus the priors $\pi(\theta) \propto \theta_2^3$ and $\pi(\theta) \propto \theta_1^3/\theta_2$ also satisfy (4.3.9) in Examples 4.3.4 and 4.3.5 respectively. Similarly, in Example 4.3.7, it can be seen that any prior of the form

$$\pi(\theta) \propto (\theta_1^{r_1}\theta_2^{r_2})^{-1}, \tag{4.3.14}$$

where r_1 and r_2 are real numbers, satisfies (4.3.9) if and only if

$$3 - r_2 - r_1^2 - r_2^2 = 0. \tag{4.3.15}$$

Clearly, (4.3.15) yields an infinite class of HPD matching priors including the one that was previously mentioned in Example 4.3.7. These alternative solutions to (4.3.9) available in the examples, except the prior $\pi(\theta) \propto \theta_1^3/\theta_2$ in Example 4.3.5, also ensure the propriety of the posteriors in the sense indicated above.

We now discuss the issue of invariance of HPD matching priors. To motivate the ideas, Example 4.3.7 is revisited. Suppose the model (4.3.13) for this example is parameterized via $\mu = (\mu_1, \mu_2)^T$ and consider HPD matching priors for μ. Will the transformed versions of these priors, under the $\theta-$parameterization, be HPD matching for θ as well? By (4.3.13), under the $\mu-$parameterization,

$$I_{11} = \mu_1^{-2}, \quad I_{22} = \mu_2^{-2}, \quad I_{12} = 0,$$

$$L_{111} = 4\mu_1^{-3}, \quad L_{112} = L_{122} = 0, \quad L_{222} = 4\mu_2^{-3}.$$

Hence by (4.3.9), any prior of the form

$$\pi^*(\mu) \propto (\mu_1^{s_1}\mu_2^{s_2})^{-1}, \tag{4.3.16}$$

where s_1 and s_2 are real numbers, is HPD matching for μ if and only if

$$4 - s_1 - s_2 - s_1^2 - s_2^2 = 0. \tag{4.3.17}$$

Since $\mu_1 = (\theta_1\theta_2)^{1/2}$ and $\mu_2 = (\theta_2/\theta_1)^{1/2}$, it is easy to see that the transformed version of (4.3.16) under the $\theta-$parameterization is of the form (4.3.14) with

$$r_1 = \frac{1}{2}(s_1 - s_2) + 1 , \quad r_2 = \frac{1}{2}(s_1 + s_2) . \qquad (4.3.18)$$

Employing (4.3.18) in (4.3.15), on simplification, such a transformed version is HPD matching for θ if and only if

$$4 - 3s_1 + s_2 - s_1^2 - s_2^2 = 0 . \qquad (4.3.19)$$

Note that (4.3.17) and (4.3.19) are not equivalent and that a choice of (s_1, s_2) satisfying the HPD matching condition (4.3.17) for μ also satisfies the condition (4.3.19) for θ if and only if either $s_1 = s_2 = 1$ or $s_1 = s_2 = -2$. Thus HPD matching priors for μ do not necessarily transform to HPD matching priors for θ - e.g., the choice $s_1 = 1$, $s_2 = -2$ satisfies (4.3.17) but not (4.3.19).

More generally, suppose a model has two alternative parameterizations, say via $\lambda = (\lambda_1, \ldots, \lambda_p)^T$ and $\theta = (\theta_1, \ldots, \theta_p)^T$, where $\theta = g(\lambda)$ is a one-to-one transformation of λ with a nonsingular Jacobian matrix for all λ. Then from the above example it is clear that an HPD matching for λ under the $\lambda-$parameterization is not guaranteed to transform to a prior enjoying the same property for θ under the $\theta-$parameterization. This phenomenon is commonly known as the lack of invariance of HPD matching priors. A closer examination, however, reveals that this lack of invariance is actually far less disturbing than it appears to be at the first sight. This becomes evident as soon as one realizes that in the above the object of interest also varies over the two parameterizations: it is λ under the $\lambda-$parameterization and $\theta(\equiv g(\lambda))$ under the $\theta-$parameterization. Indeed, if the same object of interest is considered then one is relieved to note that HPD matching priors are invariant of the parameterization adopted.

For further elucidation of the point just mentioned, continuing with the setup of the last paragraph, suppose the same object of interest, say β, is considered under both the parameterizations, where β equals θ or $g(\lambda)$ under the $\theta-$ or $\lambda-$parameterizations respectively. Let $\pi^*(\lambda)$ and $\pi(\theta)$ be priors, under the two parameterizations, that are transformed versions of each other. Since $\theta = g(\lambda)$, it is easy to see that the posterior density of β under $\pi^*(\lambda)$ and the $\lambda-$parameterization is the same as that under $\pi(\theta)$ and the $\theta-$parameterization. Hence denoting this posterior density by $\tilde{\pi}(\beta|X)$, for any observational function $q(X)$ one gets

$$P^\pi\{\tilde{\pi}(\beta|X) \geq q(X)|X\}$$
$$= \int_{\{\beta:\tilde{\pi}(\beta|X)\geq q(X)\}} \tilde{\pi}(\beta|X)\mathrm{d}\beta$$
$$= P^{\pi^*}\{\tilde{\pi}(\beta|X) \geq q(X)|X\} , \qquad (4.3.20)$$

where, as usual, $P^\pi\{\cdot|X\}$ and $P^{\pi^*}\{\cdot|X\}$ are the posterior probability measures under $\pi(\theta)$ and $\pi^*(\lambda)$ respectively. Similarly, with $\theta = g(\lambda)$, from the definition of β,

$$
\begin{aligned}
& P_\theta\{\widetilde{\pi}(\beta|X) \geq q(X)\} \\
={} & P_\theta\{\widetilde{\pi}(\theta|X) \geq q(X)\} \\
={} & P_\lambda\{\widetilde{\pi}(g(\lambda)|X) \geq q(X)\} \\
={} & P_\lambda\{\widetilde{\pi}(\beta|X) \geq q(X)\}\,.
\end{aligned}
\tag{4.3.21}
$$

Equations (4.3.20) and (4.3.21) reveal that $\pi^*(\lambda)$ is HPD matching for $\beta(= g(\lambda))$ under the $\lambda-$parameterization if and only if $\pi(\theta)$ is so for $\beta(= \theta)$ under the $\theta-$parameterization. Since $\pi^*(\lambda)$ and $\pi(\theta)$ are transformed versions of each other, one reaches the satisfying conclusion that the problem of finding an HPD matching prior is invariant of the parameterization adopted as long as the object of interest is not altered.

4.4 Results in the presence of nuisance parameters

Continuing with the setup and assumptions of Section 2.2, now suppose interest lies in θ_1 and let $\theta_2,\ldots,\theta_p$ be nuisance parameters. As before, an HPD matching prior for θ_1 is defined as one that ensures frequentist validity of HPD regions for θ_1 with margin of error $o(n^{-1})$. In order to facilitate the presentation of the results without making the notation too heavy, we work under an orthogonal parameterization, i.e., assume that $I_{1j} = 0\ (2 \leq j \leq p)$, identically in θ. Since θ_1 is one-dimensional, as noted in subsection 2.5.3, this can always be achieved by suitably choosing the nuisance parameters $\theta_2,\ldots,\theta_p$.

Theorem 4.4.1. *Suppose orthogonal parameterization holds. Then a prior $\pi(\cdot)$ is HPD matching for θ_1 if and only if it satisfies the partial differential equation*

$$
\sum_{s=2}^{p}\sum_{u=2}^{p} D_u\{\pi(\theta)I_{11}^{-1}I^{su}L_{11s}\} + D_1\{\pi(\theta)I_{11}^{-2}L_{111}\} - D_1^2\{\pi(\theta)I_{11}^{-1}\} = 0\,.
$$

$$
\tag{4.4.1}
$$

♣

Theorem 4.4.1 can be proved along the line of Theorem 4.3.1 with considerable additional algebra. The proof for the case $p = 2$ is available in Ghosh and Mukerjee (1995a). The case of general p is similar and the details are omitted here. One may wonder about the possible existence of a more general version of Theorem 4.4.1 that is valid even without an orthogonal parameterization. Such a version is potentially useful in situations where the explicit determination of an orthogonal parameterization is difficult. Later in this section, we touch upon this point.

The setting of Theorem 4.4.1 is similar to that of subsection 2.5.3 where first and second order matching priors for posterior quantiles of θ_1 were characterized under an orthogonal parameterization. We, therefore, consider it appropriate to present a comparison with the results reported there. To that effect, observe that (4.4.1) is similar to, but not identical with, the second order matching condition (2.5.24) for posterior quantiles. From subsection 2.5.3, now recall that a prior $\pi(\cdot)$ is first order matching for posterior quantiles of θ_1, under an orthogonal parameterization, if and only if it is of the form $\pi(\theta) = d(\theta^{(2)})I_{11}^{1/2}$, where $d(\cdot)(> 0)$ is any smooth function of $\theta^{(2)} = (\theta_2, \ldots, \theta_p)^T$. An algebra similar to the derivation of (2.5.26) shows that a prior of this kind satisfies (4.4.1) if and only if

$$\sum_{s=2}^{p} \sum_{u=2}^{p} D_u\{d(\theta^{(2)})I_{11}^{-1/2}I^{su}L_{11s}\} + \frac{1}{2}d(\theta^{(2)})D_1\{I_{11}^{-3/2}(L_{111} - L_{1,11})\} = 0 .$$

(4.4.2)

The above is again similar to, but not generally identical with, the corresponding second order matching condition (2.5.26) for posterior quantiles. From the regularity condition (2.5.5), it can be easily checked (cf. Datta, Ghosh and Mukerjee, 2000) that the conditions (2.5.26) and (4.4.2) become equivalent in the special case of models satisfying

$$D_1(I_{11}^{-3/2}L_{111}) = 0 .$$

(4.4.3)

Thus if (4.4.3) holds then, under an orthogonal parameterization, any second order matching prior for posterior quantiles of θ_1 is also HPD matching for θ_1.

For the purpose of illustration, we now revisit several examples. In many of these examples, the object of interest is originally a parametric function which is reduced to a canonical interest parameter θ_1 via reparameterization. This is justified since, as in the last section, HPD matching priors are invariant of the parameterization as long as the object of interest, viewed either as a parametric function under an original parameterization or as a canonical parameter after reparameterization, remains unaltered. Furthermore, when one works with a canonical interest parameter θ_1, the same argument entails invariance of HPD matching priors with respect to the choice of the nuisance parameters $\theta_2, \ldots, \theta_p$. This, in turn, legitimizes working with an orthogonal parameterization, as done here, from the point of view of invariance.

In Examples 2.5.3, 2.5.4, 2.6.3, 2.6.5, 2.6.6 and 2.6.8, orthogonal parameterization holds and we have

$$I_{11} = \theta_3/\theta_2 , \quad L_{111} = 0 ,$$

$$I_{11} \propto \theta_1^{-2} , \quad L_{111} \propto \theta_1^{-3} ,$$

$$I_{11} = (\theta_3^2 + \theta_4^2)^{-1} , \quad L_{111} = 0 ,$$

$$I_{11} = \sum_{j=1}^{t} z_j^2 , \quad L_{111} = \sum_{j=1}^{t} z_j^3 ,$$

$$I_{11} \propto \theta_1^{-2} , \quad L_{111} \propto \theta_1^{-3} ,$$

and

$$I_{11} \propto (1 + t\theta_1)^{-2} , \quad L_{111} \propto (1 + t\theta_1)^{-3} ,$$

respectively. Hence, in each of these examples, (4.4.3) is satisfied and any second order matching prior for posterior quantiles of θ_1 is HPD matching as well for θ_1. By (2.5.17), it is evident that the same conclusion holds also for location-scale models satisfying orthogonal parameterization, whether θ_1 or θ_2 is the parameter of interest.

We next consider several examples where (4.4.3) is not met.

Example 4.4.1. (Ghosh and Mukerjee, 1995a) We revisit Example 2.6.1 where interest lies in the ratio of independent normal means. Here orthogonal parameterization holds. Furthermore,

$$I_{11} = \theta_2^2/(\theta_1^2 + 1)^2 , \quad I_{22} = 1 , \quad L_{112} = -\theta_2/(\theta_1^2 + 1)^2 ,$$

$$L_{111} = -3L_{1,11} = 6\theta_1\theta_2^2/(\theta_1^2 + 1)^3 .$$

Evidently (4.4.3) does not hold. In addition, (4.4.2) reduces to

$$\theta_2 D_2\{d(\theta_2)\} = 4(\theta_1^2 + 1)d(\theta_2) ,$$

which does not admit any solution for $d(\theta_2)$. Thus no first order matching prior for posterior quantiles of θ_1 is HPD matching for θ_1. Solutions to the HPD matching condition (4.4.1) for θ_1, however, exist. For example, it can be seen that any prior of the form

$$\pi(\theta) \propto \theta_1^{r_1}\theta_2^{r_2}/(\theta_1^2 + 1)^{r_3} ,$$

where r_1, r_2 and r_3 are real, satisfies (4.4.1) if and only if one of the following holds:

(a) $r_1 = 0$, $r_2 = 6$, $r_3 = 3/2$,
(b) $r_1 = 1$, $r_2 = 13$, $r_3 = 2$,
(c) $r_1 = 0$, $r_2 = 1$, $r_3 = -1$,
(d) $r_1 = 1$, $r_2 = -2$, $r_3 = -1/2$. ♣

Example 4.4.2. We revisit Example 2.6.2 where interest lies in the product of independent normal means. Here orthogonal parameterization holds and

$$I_{11} = 4I_{22} = (4\theta_1^2 + \theta_2^2)^{-1/2} , \quad L_{112} = \frac{1}{2}\theta_2(4\theta_1^2 + \theta_2^2)^{-3/2} ,$$

$$L_{111} = -3L_{1,11} = 6\theta_1(4\theta_1^2 + \theta_2^2)^{-3/2} .$$

Again, (4.4.3) does not hold and one can check that (4.4.2) has no solution for $d(\theta_2)$, the implication of this being as in the last example. On the other hand, (4.4.1) admits a solution, namely, $\pi(\theta) \propto (4\theta_1^2 + \theta_2^2)^{1/2}$. ♣

Example 4.4.3. We revisit Example 4.3.5. Again, orthogonal parameterization holds. Also,

$$I_{11} = \theta_1^{-3}\theta_2 , \quad I_{22} = \frac{1}{2}\theta_2^{-2} , \quad L_{112} = -\theta_1^{-3} ,$$

$$L_{111} = -2L_{1,11} = 6\theta_1^{-4}\theta_2 .$$

Thus, in this example too, (4.4.3) does not hold and (4.4.2) has no solution for $d(\theta_2)$. However, solutions to (4.4.1) exist. One such solution is given by $\pi(\theta) \propto (\theta_1^2\theta_2)^{-1}$. Interestingly, as seen in Example 4.3.5, this prior is HPD matching also when interest lies in $\theta = (\theta_1, \theta_2)^T$.   ♣

It can be seen that the solutions to (4.4.1), as reported in the last two examples, entail the propriety of the posteriors, with P_θ-probability unity for all θ, whenever n is sufficiently large. The same remark holds for the solutions corresponding to (a) and (b) in Example 4.4.1. The solutions given by (c) and (d) in the same example, on the other hand, fail to ensure the propriety of the posterior even for large n and are hence useless.

DiCiccio and Stern (1994) and Ghosh and Mukerjee (1995a) gave very general results characterizing HPD matching priors when both the interest and the nuisance parameters are possibly multidimensional and no assumption about orthogonal parameterization is made. A more general version of Theorem 4.4.1, valid even without an orthogonal parameterization, can be deduced from these results. Since these developments involve somewhat complicated notation, we do not discuss them here but refer to the original sources. We also refer to Rousseau (2002) for a study of HPD matching priors in the discrete case via continuity corrections.

While concluding this chapter, we remark that HPD regions with approximate frequentist validity tend to have an edge over other confidence sets under the purely frequentist criterion of expected volume. The interested reader may see Mukerjee and Reid (1999b) and Datta and DiCiccio (2001) in this connection.

Matching Priors for Other Credible Regions

5.1 Introduction

In this chapter, we focus on posterior credible regions obtained by the inversion of certain commonly used statistics. Priors ensuring approximate frequentist validity of such regions are characterized. This is done with margin of error $o(n^{-1})$, where n is the sample size. The results, when combined with those of the previous chapter, can help in narrowing down the choice of matching priors especially when the interest parameter is multidimensional.

We begin by finding, in the next section, an explicit form of a credible region given by the inversion of the likelihood ratio (LR) statistic. The corresponding matching priors are characterized in the same section. These developments are quite similar to those in the last chapter for an HPD region. An expression for the frequentist Bartlett adjustment to the LR statistic follows as a by-product. This is discussed in Section 5.3. Matching priors associated with credible regions given by the inversion of posterior versions of Rao's score and Wald's statistics are characterized in Section 5.4. A region given by the latter statistic is also known as an ellipsoidal region. Finally, in Section 5.5, we address a related problem of finding for a given prior perturbed ellipsoidal and HPD regions so as to achieve both Bayesian and frequentist validity with margin of error $o(n^{-1})$.

Throughout this chapter, we continue with the setup and assumptions of Section 2.2. Unless otherwise specified, the notation is also as described there.

5.2 Matching priors associated with the LR statistic

5.2.1 Credible region via the LR statistic

Suppose interest lies in the entire parameter vector $\theta = (\theta_1, \ldots, \theta_p)^T$, i.e., there is no nuisance parameter. The LR statistic for θ is given by

$$M_{\text{LR}}(\theta, X) = 2n\{\ell(\widehat{\theta}) - \ell(\theta)\}, \qquad (5.2.1)$$

where $X = (X_1, \ldots, X_n)^T$ and, as usual, $\ell(\theta) = n^{-1}\sum_{i=1}^{n}\log f(X_i;\theta)$. Given a prior $\pi(\cdot)$, the inversion of (5.2.1) yields a posterior credible region for θ as

$$Q_{\text{LR}}^{(1-\alpha)}(\pi, X) = \{\theta : M_{\text{LR}}(\theta, X) \le k_{1-\alpha}(\pi, X)\}, \qquad (5.2.2)$$

where $k_{1-\alpha}(\pi, X)$, which may depend on $\pi(\cdot)$ and X but not on θ, has to be so chosen that the relation

$$P^\pi\{\theta \in Q_{\text{LR}}^{(1-\alpha)}(\pi, X)|X\} = 1 - \alpha + o(n^{-1}) \qquad (5.2.3)$$

holds.

Note that (5.2.2) and (5.2.3) are analogous to (4.2.1) and (4.2.4) respectively. Therefore, as in Section 4.2, consideration of the approximate posterior characteristic function of $M_{\text{LR}}(\theta, X)$ facilitates the explicit determination of $k_{1-\alpha}(\pi, X)$ and hence that of $Q_{\text{LR}}^{(1-\alpha)}(\pi, X)$. Let $\xi = (-1)^{1/2}t$, where t is an auxiliary variable, and define $W_1, \ldots, W_4$ as in (2.2.18). Then, analogously to Lemma 4.2.1, the following result holds.

Lemma 5.2.1. *Under the prior $\pi(\cdot)$, the approximate posterior characteristic function of $M_{LR}(\theta, X)$ is given by*

$$E^\pi[\exp\{\xi M_{LR}(\theta, X)\}|X]$$
$$= (1 - 2\xi)^{-\frac{1}{2}p}\left\{1 + \frac{1}{2}n^{-1}\left(\frac{1}{1 - 2\xi} - 1\right)\left(W_1 + \frac{1}{3}W_2 + \frac{1}{12}W_3 + \frac{1}{36}W_4\right)\right\}$$
$$+ o(n^{-1}).$$

Proof. Let $R_1(h), \ldots, R_4(h)$ be as shown in (2.2.12), where $h = n^{1/2}(\theta - \widehat{\theta})$. By (2.2.11) and (5.2.1),

$$M_{\text{LR}}(\theta, X) = h^T Ch - \frac{1}{3}n^{-1/2}R_3(h) - \frac{1}{12}n^{-1}R_4(h) + o(n^{-1}). \qquad (5.2.4)$$

Hence by (2.2.19), after a little algebra,

$$E^\pi[\exp\{\xi M_{\text{LR}}(\theta, X)\}|X]$$
$$= \int [\exp\{\xi M_{\text{LR}}(\theta, X)\}]\pi^*(h|X)dh$$
$$= \int \left[1 + n^{-1/2}\left\{R_1(h) + \frac{1}{6}(1 - 2\xi)R_3(h)\right\}\right.$$
$$+ n^{-1}\left\{\frac{1}{2}R_2(h) + \frac{1}{72}(1 - 2\xi)^2\left(R_3(h)\right)^2\right.$$
$$+ \frac{1}{6}(1 - 2\xi)R_1(h)R_3(h) + \frac{1}{24}(1 - 2\xi)R_4(h)$$
$$\left.\left. - \frac{1}{2}W_1 - \frac{1}{6}W_2 - \frac{1}{24}W_3 - \frac{1}{72}W_4\right\}\right]e^{\xi h^T Ch}\phi_p(h; C^{-1})dh + o(n^{-1}).$$

The proof can now be completed proceeding exactly as in Lemma 4.2.1.   ♣

Let z^2 be the $(1 - \alpha)$th quantile of the chi-square distribution with p degrees of freedom. Inversion of the approximate posterior characteristic function given in Lemma 5.2.1 shows that if

$$k_{1-\alpha}(\pi, X) = z^2\left\{1 + (np)^{-1}\left(W_1 + \frac{1}{3}W_2 + \frac{1}{12}W_3 + \frac{1}{36}W_4\right)\right\} \qquad (5.2.5)$$

then (5.2.3) holds. This again follows along the line of Section 4.2. Thus the credible region $Q_{\mathrm{LR}}^{(1-\alpha)}(\pi, X)$ is explicitly described by (5.2.2) and (5.2.5).

5.2.2 Matching priors

We now execute Steps 1–3 of Section 1.2 in order to calculate the frequentist coverage probability $P_\theta\{\theta \in Q_{\mathrm{LR}}^{(1-\alpha)}(\pi, X)\}$, up to the order of approximation $o(n^{-1})$.

Step 1: We first consider $P^{\overline{\pi}}\{\theta \in Q_{\mathrm{LR}}^{(1-\alpha)}(\pi, X)|X\}$, where the auxiliary prior $\overline{\pi}(\cdot)$, satisfying the conditions in Bickel and Ghosh (1990), is as described in Section 2.4. By Lemma 5.2.1, with $\pi(\cdot)$ there replaced by $\overline{\pi}(\cdot)$, the approximate posterior characteristic function of $M_{\mathrm{LR}}(\theta, X)$ under $\overline{\pi}(\cdot)$ is given by

$$
\begin{aligned}
&E^{\overline{\pi}}[\exp\{\xi M_{\mathrm{LR}}(\theta, X)\}|X] \\
&= (1 - 2\xi)^{-\frac{1}{2}p}\left\{1 + \frac{1}{2}n^{-1}\left(\frac{1}{1 - 2\xi} - 1\right)\left(\overline{W}_1 + \frac{1}{3}\overline{W}_2 + \frac{1}{12}W_3 + \frac{1}{36}W_4\right)\right\} \\
&\quad + o(n^{-1}),
\end{aligned} \qquad (5.2.6)
$$

where $\overline{W}_1$ and $\overline{W}_2$, which are counterparts of W_1 and W_2 under $\overline{\pi}(\cdot)$, are as shown in (4.3.3); note that W_3 and W_4 do not depend on the prior. By (5.2.2) and (5.2.5), proceeding as in the derivation of (4.3.4), the inversion of (5.2.6) yields

$$
\begin{aligned}
&P^{\overline{\pi}}\{\theta \in Q_{\mathrm{LR}}^{(1-\alpha)}(\pi, X)|X\} \\
&= 1 - \alpha + (np)^{-1}z^2\psi_p(z^2)\left\{W_1 - \overline{W}_1 + \frac{1}{3}\left(W_2 - \overline{W}_2\right)\right\} \\
&\quad + o(n^{-1}),
\end{aligned} \qquad (5.2.7)
$$

$\psi_p(\cdot)$ being the chi-square density with p degrees of freedom.
Step 2: From (5.2.7), analogously to (4.3.5), one gets

$$
\begin{aligned}
&E_\theta[P^{\overline{\pi}}\{\theta \in Q_{\mathrm{LR}}^{(1-\alpha)}(\pi, X)|X\}] \\
&= 1 - \alpha + (np)^{-1}z^2\psi_p(z^2)\left\{I^{jr}\left(\frac{\pi_{jr}(\theta)}{\pi(\theta)} - \frac{\overline{\pi}_{jr}(\theta)}{\overline{\pi}(\theta)}\right)\right. \\
&\quad \left. + L_{jrs}I^{jr}I^{su}\left(\frac{\pi_u(\theta)}{\pi(\theta)} - \frac{\overline{\pi}_u(\theta)}{\overline{\pi}(\theta)}\right)\right\} + o(n^{-1}),
\end{aligned} \qquad (5.2.8)
$$

for θ in the interior of the support of $\overline{\pi}(\cdot)$.

Step 3: We proceed again as in the corresponding step in Section 2.4. By (5.2.8), this yields

$$P_\theta\{\theta \in Q_{\mathrm{LR}}^{(1-\alpha)}(\pi, X)\} = 1 - \alpha + (np)^{-1}\frac{z^2\psi_p(z^2)}{\pi(\theta)}\Delta(\pi,\theta) + o(n^{-1}) , \quad (5.2.9)$$

for all θ, where

$$\begin{aligned}
\Delta(\pi,\theta) &= I^{jr}\pi_{jr}(\theta) - \pi(\theta)D_jD_r(I^{jr}) \\
&\quad + L_{jrs}I^{jr}I^{su}\pi_u(\theta) + \pi(\theta)D_u(L_{jrs}I^{jr}I^{su}) \\
&= D_u\{\pi(\theta)L_{jrs}I^{jr}I^{su}\} - D_jD_r\{\pi(\theta)I^{jr}\} \\
&\quad + 2D_r\{I^{jr}\pi_j(\theta)\} , \quad (5.2.10)
\end{aligned}$$

using (4.3.8).

We are now in a position to characterize priors that ensure

$$P_\theta\{\theta \in Q_{\mathrm{LR}}^{(1-\alpha)}(\pi, X)\} = 1 - \alpha + o(n^{-1}) ,$$

for all α and θ. These are called matching priors for credible regions for θ as given by the inversion of the LR statistic or, briefly, LR matching priors for θ. From (5.2.9) and (5.2.10), the following result is evident.

Theorem 5.2.1. *A prior $\pi(\cdot)$ is LR matching for θ if and only if it satisfies the partial differential equation*

$$D_u\{\pi(\theta)L_{jrs}I^{jr}I^{su}\} - D_jD_r\{\pi(\theta)I^{jr}\} + 2D_r\{I^{jr}\pi_j(\theta)\} = 0 . \quad (5.2.11)$$

♣

The above result is due to Ghosh and Mukerjee (1991) who reported it in another equivalent form. Observe that (5.2.11) is similar to, but not identical with, the corresponding matching condition (4.3.9) for HPD regions. In particular, if $p = 1$ then $\theta = \theta_1$, I is a scalar and (5.2.11) becomes

$$\frac{d}{d\theta}\{\pi(\theta)L_{111}I^{-2}\} - \frac{d^2}{d\theta^2}\{\pi(\theta)I^{-1}\} + 2\frac{d}{d\theta}\{I^{-1}\Big(\frac{d\pi(\theta)}{d\theta}\Big)\} = 0 ,$$

i.e.,

$$\pi(\theta)L_{111}I^{-2} - \frac{d}{d\theta}\{\pi(\theta)I^{-1}\} + 2I^{-1}\Big(\frac{d\pi(\theta)}{d\theta}\Big) = \text{constant} .$$

In view of the regularity condition (2.5.4), the above is equivalent to

$$I^{-1}\Big(\frac{d\pi(\theta)}{d\theta}\Big) - \pi(\theta)L_{1,11}I^{-2} = \text{constant} , \quad (5.2.12)$$

which again is similar to, though not identical with, the corresponding condition (4.3.10) for HPD regions. Equation (5.2.12) is in agreement with the findings of Severini (1991) who studied this problem for scalar θ.

Continuing with $p = 1$, as in Section 4.3, it is of interest to examine when a prior of the form $\pi(\theta) \propto I^r$, where r is a real number, satisfies (5.2.12). Using (2.5.4) again, (5.2.12) holds for such a prior if and only if

$$I^{r-2}\{rL_{111} + (r+1)L_{1,11}\} = \text{constant} . \tag{5.2.13}$$

Taking $r = 1/2$ in the above and invoking the regularity condition (2.5.5), it follows that Jeffreys' prior satisfies (5.2.12) if and only if

$$I^{-3/2}L_{1,1,1} = \text{constant} . \tag{5.2.14}$$

Interestingly, by Theorem 2.5.1(b), the above is also the condition under which Jeffreys' prior is second order matching for the posterior quantiles of θ.

As noted in subsection 2.5.1, (5.2.14) holds for the one-parameter location or scale models. It also holds for the model considered in Example 4.3.1. Hence in these situations, Jeffreys' prior is LR matching for θ. On the other hand, for the bivariate normal model with unknown correlation coefficient as considered in Examples 2.5.2 and 4.3.2, the condition (5.2.14) is not met but (5.2.13) holds with $r = 1$. Thus for this model, $\pi(\theta) \propto I$ is LR matching for θ though Jeffreys' prior does not enjoy this property.

With a view to illustrating the LR matching condition (5.2.11) applicable to general θ, we now revisit the other examples of Section 4.3. Observe that the left-hand side of (5.2.11) incorporates that of the HPD matching condition (4.3.9) and includes an additional term $2D_r\{I^{jr}\pi_j(\theta)\}$. Hence an HPD matching prior $\pi(\cdot)$ satisfying (4.3.9) is also LR matching for θ if and only if it satisfies

$$D_r\{I^{jr}\pi_j(\theta)\} = 0 \tag{5.2.15}$$

as well. This happens for the priors

$$\pi(\theta) \propto \theta_2^{-1} , \quad \pi(\theta) \propto \theta_2^{-1} , \quad \pi(\theta) \propto (\theta_1^2\theta_2)^{-1} , \quad \pi(\theta) \propto (\theta_1\theta_2)^{-1} \tag{5.2.16}$$

reported in Examples 4.3.3, 4.3.4, 4.3.5 and 4.3.7 respectively. Thus these priors are both HPD and LR matching for θ in the respective examples. On the other hand, the prior $\pi(\theta) \propto (\theta_1^3\theta_2)^{-1}$ obtained in Example 4.3.6 does not satisfy (5.2.15). Therefore, it is not LR matching for θ despite being HPD matching. In the setup of this example, however, it can be seen that the prior $\pi(\theta) \propto \theta_1/\theta_2$ satisfies (5.2.11) and is hence LR matching for θ. One can also check that this prior as well as the LR matching prior $\pi(\theta) \propto I$, reported in the last paragraph with reference to Example 4.3.2, entail propriety of the posteriors, with P_θ−probability unity for all θ, whenever n is sufficiently large. As noted in Section 4.3, the same conclusion holds also for the priors listed in (5.2.16).

The LR matching condition (5.2.11), when applied in conjunction with the HPD matching condition (4.3.9), can substantially narrow down the choice of matching priors especially for multidimensional θ. An illustrative example follows.

Example 5.2.1. We revisit Example 4.3.7 and consider priors of the form $\pi(\theta) \propto (\theta_1^{r_1} \theta_2^{r_2})^{-1}$, where r_1 and r_2 are real numbers. Using the expressions for the I_{jr} and the L_{jrs} as shown in Example 4.3.7, it is not hard to see that any such prior satisfies the LR matching condition (5.2.11) if and only if

$$3 - 2r_1 - 3r_2 + r_1^2 + r_2^2 = 0 . \qquad (5.2.17)$$

On the other hand, as noted in Section 4.3, any prior of this kind satisfies the HPD matching condition (4.3.9) if and only if (4.3.15) holds. Thus if one insists on both LR and HPD matching properties for θ then the pair (r_1, r_2) must satisfy both (5.2.17) and (4.3.15). The only choices of (r_1, r_2) meeting both these conditions are (1,1) and (3/5, 6/5). Hence consideration of (5.2.17) significantly narrows down the class of solutions to (4.3.15).   ♣

5.2.3 Results in the presence of nuisance parameters

The LR matching condition (5.2.11) derived in the last subsection allows θ to be possibly multidimensional but presumes the absence of any nuisance parameter. Several results on the characterization of LR matching priors in the presence of nuisance parameters have also been reported in the literature. The most comprehensive work in this direction is due to DiCiccio and Stern (1994) who allowed both the interest and nuisance parameters to be possibly multi-dimensional and made no assumption regarding orthogonal parameterization. Their results yield matching conditions not only for the usual LR statistic but also for similar statistics arising from modified versions of the profile likelihood, such as the conditional likelihood of Cox and Reid (1987) and the adjusted likelihood of McCullagh and Tibshirani (1990). These modifications have been proposed with a view to neutralizing, in various senses, the presence of unknown nuisance parameters. Earlier, Ghosh and Mukerjee (1992b, 1994b) derived special cases of the aforesaid matching conditions for the situation where both the interest and nuisance parameters are one-dimensional and orthogonal parameterization holds. Yin and Ghosh (1997) studied the connection between second order matching conditions for posterior quantiles (vide subsection 2.5.3) and a matching condition associated with the LR statistic arising from Cox and Reid's (1987) conditional likelihood. They also extended the results of Ghosh and Mukerjee (1992b) to the case of a possibly multidimensional nuisance parameter. We refer to the original papers for details but, in order to give a flavor of these developments, quote and illustrate one of the findings in Yin and Ghosh (1997).

Let θ_1 be the interest parameter and $\theta_2, \ldots, \theta_p$ be the nuisance parameters. Suppose orthogonal parameterization holds, i.e., $I_{1j} = 0$ $(2 \leq j \leq p)$, identically in θ. Let $\theta^{(2)} = (\theta_2, \ldots, \theta_p)^T$ and write $\widehat{\theta}^{(2)}(\theta_1) = (\widehat{\theta}_2(\theta_1), \cdots, \widehat{\theta}_p(\theta_1))^T$ for the maximum likelihood estimator of $\theta^{(2)}$ given θ_1. In the same way as (5.2.1), the usual LR statistic for θ_1 is given by

$$M_{LR}^*(\theta_1, X) = 2n\{\ell(\widehat{\theta}) - \ell^*(\theta_1)\},$$

where $\ell^*(\theta_1)$ equals $\ell(\theta)$ as evaluated at $\theta = (\theta_1, \widehat{\theta}_2(\theta_1), \ldots, \widehat{\theta}_p(\theta_1))^T$. As before, an LR matching prior for θ_1 is defined as one that ensures frequentist validity, with margin of error $o(n^{-1})$, of posterior credible regions for θ_1 given by the inversion of $M_{LR}^*(\theta_1, X)$. Then the following result, due to Yin and Ghosh (1997), holds.

Theorem 5.2.2. *Suppose orthogonal parameterization holds. Then a prior $\pi(\cdot)$ is LR matching for θ_1 if and only if it satisfies the partial differential equation*

$$\sum_{s=2}^{p}\sum_{u=2}^{p} D_u\{\pi(\theta)I_{11}^{-1}I^{su}L_{11s}\}$$

$$+ D_1[I_{11}^{-1}\{\pi_1(\theta) - \pi(\theta)(I_{11}^{-1}L_{1,11} - \sum_{s=2}^{p}\sum_{u=2}^{p}I^{su}L_{1su})\}] = 0. (5.2.18)$$

♣

As noted in subsection 2.5.3, a prior is first order matching for posterior quantiles of θ_1, under an orthogonal parameterization, if and only if it is of the form $\pi(\theta) = d(\theta^{(2)})I_{11}^{1/2}$, where $d(\cdot)(> 0)$ is any smooth function of $\theta^{(2)}$. Using the regularity conditions (2.5.5) and

$$D_1 I_{11} = -(L_{1,11} + L_{111}), \tag{5.2.19}$$

which is analogous to (2.5.4), a prior of this kind satisfies (5.2.18) if and only if

$$\sum_{s=2}^{p}\sum_{u=2}^{p} D_u\{d(\theta^{(2)})I_{11}^{-1/2}I^{su}L_{11s}\}$$

$$+ d(\theta^{(2)})D_1\left(\frac{1}{2}I_{11}^{-3/2}L_{1,1,1} + I_{11}^{-1/2}\sum_{s=2}^{p}\sum_{u=2}^{p}I^{su}L_{1su}\right) = 0. \tag{5.2.20}$$

The above is similar to, though not generally identical with, the corresponding second order matching condition (2.5.26) for posterior quantiles of θ_1. It is easily seen that the conditions (2.5.26) and (5.2.20) become equivalent in the special case of models satisfying

$$D_1\left(\frac{1}{3}I_{11}^{-3/2}L_{1,1,1} + I_{11}^{-1/2}\sum_{s=2}^{p}\sum_{u=2}^{p}I^{su}L_{1su}\right) = 0. \tag{5.2.21}$$

Clearly, if (5.2.21) holds then, under an orthogonal parameterization, any second order matching prior for the posterior quantiles of θ_1 is also LR matching for θ_1.

For illustration, we revisit Examples 2.5.4, 2.6.1 and 2.6.5. In each of these examples, orthogonal parameterization holds. In the first two of these, $L_{1,1,1} = L_{122} = 0$, while in the third one each of I_{11}, I_{22}, $L_{1,1,1}$ and L_{122} is free from θ_1. Hence (5.2.21) is met in all these examples. Consequently, the unique second order matching priors for the posterior quantiles of θ_1, as obtained earlier in these examples, are LR matching as well for θ_1. One can also verify that (5.2.21) holds for location-scale models satisfying orthogonal parameterization, whether the location or the scale parameter is of interest. We next consider an example where (5.2.21) is not met.

Example 5.2.2. We revisit Example 4.3.5 on the inverse Gaussian model. Here orthogonal parameterization holds. Furthermore,

$$I_{11} = \theta_1^{-3}\theta_2 , \quad I_{22} = \frac{1}{2}\theta_2^{-2} , \quad L_{112} = -\theta_1^{-3} , \quad L_{122} = 0 ,$$

$$L_{1,1,1} = -L_{1,11} = 3\theta_1^{-4}\theta_2 .$$

It is easily seen that (5.2.21) does not hold. In addition, (5.2.20) reduces to

$$\theta_2^{1/2}D_2\{d(\theta_2)\theta_2^{3/2}\} = \frac{3}{8}\theta_1 d(\theta_2) ,$$

which does not admit any solution for $d(\theta_2)$. Thus no first order matching prior for posterior quantiles of θ_1 can be LR matching for θ_1. The LR matching condition (5.2.18) for θ_1, however, has a solution, namely, $\pi(\theta) \propto (\theta_1^2\theta_2)^{-1}$. As noted in (5.2.16), this prior is LR matching also when interest lies in $\theta = (\theta_1, \theta_2)^T$. Interestingly, the findings in Examples 4.3.5 and 4.4.3 show that it is HPD matching as well when either θ or θ_1 is of interest.   ♣

For a more general comparison between the LR and HPD matching conditions (5.2.18) and (4.4.1) for θ_1, observe that the left-hand side of (4.4.1) can be expressed as

$$\sum_{s=2}^{p}\sum_{u=2}^{p}D_u\{\pi(\theta)I_{11}^{-1}I^{su}L_{11s}\} - D_1[I_{11}^{-1}\{\pi_1(\theta) + \pi(\theta)I_{11}^{-1}L_{1,11}\}] .$$

This follows using (5.2.19). Consideration of the difference of the left-hand sides of (5.2.18) and (4.4.1) now reveals that an HPD matching prior $\pi(\cdot)$ satisfying (4.4.1) is also LR matching for θ_1 if and only if it satisfies

$$D_1\left[I_{11}^{-1}\left\{2\pi_1(\theta) + \pi(\theta)\sum_{s=2}^{p}\sum_{u=2}^{p}I^{su}L_{1su}\right\}\right] = 0 . \tag{5.2.22}$$

The prior $\pi(\theta) \propto (\theta_1^2\theta_2)^{-1}$ in Example 5.2.2 indeed satisfies (5.2.22). On the other hand, (5.2.22) does not hold for any of the HPD matching priors obtained in Example 4.4.1. This, however, does not preclude the availability of an LR matching prior for θ_1 in the latter example which is in continuation of Example 2.6.1. As noted earlier in this subsection, (5.2.21) holds in this example and hence the unique second order matching prior for posterior quantiles of θ_1 is also LR matching for θ_1.

5.3 Frequentist Bartlett adjustment

It was hinted in Section 1.2 that the shrinkage argument, consisting of the Steps 1–3 discussed there, can facilitate the handling of purely frequentist problems in addition to substantially simplifying the derivation of matching priors in various contexts. As an illustration, in the spirit of Bickel and Ghosh (1990) and Ghosh and Mukerjee (1991), we now derive the frequentist Bartlett adjustment to the LR statistic. The setup is again as in subsection 5.2.1 where interest lies in the entire parameter vector θ and there is no nuisance parameter.

We begin by establishing a frequentist counterpart of Lemma 5.2.1. Analogously to (2.2.3), for $1 \leq j, r, s, u \leq p$, let

$$L_{jrsu} = E_{\theta}\{D_j D_r D_s D_u \log f(X_1; \theta)\} .$$

Lemma 5.3.1. *The approximate frequentist characteristic function of the LR statistic $M_{LR}(\theta, X)$ is given by*

$$E_{\theta}[\exp\{\xi M_{LR}(\theta, X)\}] = (1-2\xi)^{-\frac{1}{2}p}\left\{1 + \frac{1}{2}n^{-1}\left(\frac{1}{1-2\xi} - 1\right)B(\theta)\right\} + o(n^{-1}) ,$$

(5.3.1)

where

$$B(\theta) = D_j D_r(I^{jr}) - D_u(L_{jrs}I^{jr}I^{su}) + \frac{1}{4}L_{jrsu}I^{jr}I^{su}$$

$$+ \frac{1}{36}L_{jrs}L_{uvw}(9I^{jr}I^{su}I^{vw} + 6I^{ju}I^{rv}I^{sw}) .$$

(5.3.2)

Proof. Steps 1–3 of Section 1.2 are applied again with reference to the left-hand side of (5.3.1). The result of Step 1 is already shown in (5.2.6). Recalling (2.2.18) and (4.3.3), Step 2 now yields

$$E_{\theta}E^{\overline{\pi}}[\exp\{\xi M_{LR}(\theta, X)\}|X]$$
$$= (1 - 2\xi)^{-\frac{1}{2}p}\left[1 + \frac{1}{2}n^{-1}\left(\frac{1}{1-2\xi} - 1\right)\left\{\frac{I^{jr}\overline{\pi}_{jr}(\theta)}{\overline{\pi}(\theta)} + \frac{L_{jrs}I^{jr}I^{su}\overline{\pi}_u(\theta)}{\overline{\pi}(\theta)}\right.\right.$$
$$\left.\left. + \frac{1}{4}L_{jrsu}I^{jr}I^{su} + \frac{1}{36}L_{jrs}L_{uvw}(9I^{jr}I^{su}I^{vw} + 6I^{ju}I^{rv}I^{sw})\right\}\right]$$
$$+ o(n^{-1}) ,$$

for θ in the interior of the support of $\overline{\pi}(\cdot)$. The truth of (5.3.1) for all θ, with $B(\theta)$ given by (5.3.2), now follows as an immediate consequence of Step 3. ♣

Let χ_p^2 denote a chi-square random variable with p degrees of freedom. The leading term in (5.3.1) shows that $M_{LR}(\theta, X)$ is asymptotically distributed as χ_p^2, with margin of error $O(n^{-1})$, in the frequentist setup. From (5.3.1), one also gets

$$E_\theta\{M_{LR}(\theta, X)\} = p\{1 + (np)^{-1}B(\theta)\} + o(n^{-1}) \,.$$

Hence the frequentist expectation of the adjusted statistic

$$M_{LR}^{adj}(\theta, X) = M_{LR}(\theta, X)/\{1 + (np)^{-1}B(\theta)\} \qquad (5.3.3)$$

equals $p + o(n^{-1})$, i.e., this expectation equals that of χ_p^2 with margin of error $o(n^{-1})$. In Lemma 5.3.1, replacing ξ by $\xi/\{1 + (np)^{-1}B(\theta)\}$, one can also check that the approximate frequentist characteristic function of $M_{LR}^{adj}(\theta, X)$ is given by

$$E_\theta\left[\exp\left\{\xi M_{LR}^{adj}(\theta, X)\right\}\right] = (1 - 2\xi)^{-\frac{1}{2}p} + o(n^{-1}) \,.$$

This shows that $M_{LR}^{adj}(\theta, X)$ is asymptotically distributed as χ_p^2, with margin of error $o(n^{-1})$, in the frequentist setup (under wide generality, the margin of error is actually $O(n^{-2})$; see Barndorff-Nielsen and Hall, 1988). Therefore, the adjustment (5.3.3), which is based on the expectation, also entails an improved approximation to the limiting chi-square distribution. This phenomenon is known as Bartlett adjustability (Bartlett, 1937). In the frequentist setup, $M_{LR}^{adj}(\theta, X)$ is called the Bartlett adjusted version of the LR statistic and the divisor in (5.3.3), i.e., $1+(np)^{-1}B(\theta)$, is called the Bartlett adjustment factor.

Equation (5.3.2), obtained via a transparent Bayesian route, yields an explicit expression for the frequentist Bartlett adjustment factor. One can check that this is in agreement with the findings in Barndorff-Nielsen and Blæsild (1986) who considered the same problem via a direct frequentist approach in the possible presence of nuisance parameters. In particular, if $p = 1$ then $\theta = \theta_1$, I is a scalar and (5.3.2) reduces to the simple form

$$B(\theta) = \left(\frac{d^2}{d\theta^2}I^{-1}\right) - \frac{d}{d\theta}\left(L_{111}I^{-2}\right) + \frac{1}{4}L_{1111}I^{-2} + \frac{5}{12}L_{111}^2 I^{-3} \,. \qquad (5.3.4)$$

Example 5.3.1. Consider the simple exponential model given by

$$f(x; \theta) = \frac{1}{\theta}e^{-x/\theta}, \quad x > 0 \,,$$

where $\theta > 0$. Then $p = 1$, $I = \theta^{-2}$, $L_{111} = 4\theta^{-3}$, $L_{1111} = -18\theta^{-4}$, so that by (5.3.4), $B(\theta) = 1/6$. Thus the frequentist Bartlett adjustment factor is given by $1 + (6n)^{-1}$. The simple nature of this example allows one to readily check that this is in agreement with what one gets by evaluating $E_\theta\{M_{LR}(\theta, X)\}$ from first principles. ♣

5.4 Matching priors associated with Rao's score and Wald's statistics

Just as the LR statistic, two other statistics enjoy widespread popularity among statisticians. These are Rao's score and Wald's statistics (Rao, 1973, Ch. 6). Suppose interest lies in the entire parameter vector θ. Then Rao's score statistic is based on the score vector

$$\nabla \ell(\theta) = (D_1 \ell(\theta), \ldots, D_p \ell(\theta))^T .$$

In the posterior setup, with $h = (h_1, \ldots, h_p)^T = n^{1/2}(\theta - \widehat{\theta})$,

$$n^{1/2} \nabla \ell(\theta) = -Ch + o(1) , \tag{5.4.1}$$

so that by (2.2.19), $n^{1/2} \nabla \ell(\theta)$ is asymptotically p−variate normal with null mean vector and dispersion matrix C. Hence, following Rao and Mukerjee (1995), we find it natural to consider a posterior version of Rao's score statistic as given by

$$M_{\text{Rao}}(\theta, X) = n\{\nabla \ell(\theta)\}^T C^{-1}\{\nabla \ell(\theta)\} . \tag{5.4.2}$$

Similarly, Wald's statistic is based on $h = n^{1/2}(\theta - \widehat{\theta})$ which, by (2.2.19), is asymptotically p-variate normal with null mean vector and dispersion matrix C^{-1} in the posterior setup. Following Rao and Mukerjee (1995) again, this leads to a posterior version of Wald's statistic as

$$M_{\text{Wald}}(\theta, X) = n(\theta - \widehat{\theta})^T C(\theta - \widehat{\theta}) . \tag{5.4.3}$$

In analogy with (5.2.2) and (5.2.3), given a prior $\pi(\cdot)$, the inversion of (5.4.2) yields a posterior credible region for θ as

$$Q_{\text{Rao}}^{(1-\alpha)}(\pi, X) = \{\theta : M_{\text{Rao}}(\theta, X) \leq k_{1-\alpha}(\pi, X)\} , \tag{5.4.4}$$

where $k_{1-\alpha}(\pi, X)$, which may depend on $\pi(\cdot)$ and X but not on θ, has to be so chosen that the relation

$$P^\pi\{\theta \in Q_{\text{Rao}}^{(1-\alpha)}(\pi, X)|X\} = 1 - \alpha + o(n^{-1}) \tag{5.4.5}$$

holds. A credible region based on Wald's statistic can be defined similarly; in view of (5.4.3), this is also known as an ellipsoidal region. Matching conditions that entail approximate frequentist validity of these regions can be obtained along the line of Section 5.2. Key steps of the derivation, with reference to (5.4.4), are indicated below.

As a refinement to (5.4.1), for $1 \leq j \leq p$,

$$n^{1/2}D_j \ell(\theta) = -c_{jr}h_r + \frac{1}{2}n^{-1/2}a_{jrs}h_r h_s + \frac{1}{6}n^{-1}a_{jrsu}h_r h_s h_u + o(n^{-1}) .$$

Hence by (5.4.2),

$$M_{\text{Rao}}(\theta, X) = h^T C h - n^{-1/2} R_3(h)$$
$$+ n^{-1} \left\{ \frac{1}{4} c^{vw} a_{jrv} a_{suw} h_j h_r h_s h_u - \frac{1}{3} R_4(h) \right\} + o(n^{-1}) ,$$

where $R_3(h)$ and $R_4(h)$ are again as shown in (2.2.12). Proceeding as in the proofs of Lemmas 4.2.1 and 5.2.1, the approximate posterior characteristic function of $M_{\text{Rao}}(\theta, X)$ under a prior $\pi(\cdot)$ is, therefore, given by

$$E^{\pi}[\exp\{\xi M_{\text{Rao}}(\theta, X)\}|X] = (1 - 2\xi)^{-\frac{1}{2}p} \left\{ 1 + n^{-1} \sum_{j=0}^{3} H_{\text{Rao}}^{(j)} (1 - 2\xi)^{-j} \right\}$$
$$+ o(n^{-1}) , \qquad (5.4.6)$$

where

$$H_{\text{Rao}}^{(0)} = -\frac{1}{72} (36 W_1 + 12 W_2 + 3 W_3 + W_4) ,$$
$$H_{\text{Rao}}^{(1)} = \frac{1}{24} (12 W_1 + 12 W_2 + 4 W_3 + 3 W_4 - 3 W_5) ,$$
$$H_{\text{Rao}}^{(2)} = -\frac{1}{24} (8 W_2 + 3 W_3 + 4 W_4 - 3 W_5) , \quad H_{\text{Rao}}^{(3)} = \frac{1}{18} W_4 .$$

In the above, $W_1, \ldots, W_4$ are as given by (2.2.18) and

$$W_5 = c^{vw} (c^{jr} c^{su} + c^{js} c^{ru} + c^{ju} c^{rs}) a_{jrv} a_{suw} .$$

The approximate posterior characteristic function (5.4.6) is a little more involved than the corresponding expression shown in Lemma 5.2.1 for the LR statistic. Nevertheless, one can readily invert it and check that if

$$k_{1-\alpha}(\pi, X) = z^2 - n^{-1} \{\psi_p(z^2)\}^{-1} \left\{ \sum_{j=0}^{3} H_{\text{Rao}}^{(j)} \Psi_{p+2j}(z^2) \right\} , \qquad (5.4.7)$$

then (5.4.5) is satisfied. Here $\Psi_{\nu}(\cdot)$ and $\psi_{\nu}(\cdot)$ denote respectively the cumulative distribution function and density function of a chi-square variate with ν degrees of freedom ($\nu = 1, 2, \ldots$). As usual, z^2 is the $(1 - \alpha)$th quantile of the chi-square distribution with p degrees of freedom.

The shrinkage argument can now be employed to study the frequentist coverage of the credible region (5.4.4). In Step 1 of this argument, we replace $\pi(\cdot)$ by an auxiliary prior $\bar{\pi}(\cdot)$ in (5.4.6) to get

$$E^{\bar{\pi}}[\exp\{\xi M_{\text{Rao}}(\theta, X)\}|X] = (1 - 2\xi)^{-\frac{1}{2}p} \left\{ 1 + n^{-1} \sum_{j=0}^{3} \overline{H}_{\text{Rao}}^{(j)} (1 - 2\xi)^{-j} \right\}$$
$$+ o(n^{-1}) , \qquad (5.4.8)$$

where the $\overline{H}_{\text{Rao}}^{(j)}$ are obtained from the $H_{\text{Rao}}^{(j)}$ replacing W_1 and W_2 in the latter respectively by $\overline{W}_1$ and $\overline{W}_2$ as defined in (4.3.3); note that W_3, W_4 and

W_5 do not involve the prior. In view of (5.4.4) and (5.4.7), inversion of (5.4.8) yields

$$P^{\overline{\pi}}\{\theta \in Q_{\mathrm{Rao}}^{(1-\alpha)}(\pi, X)|X\}$$

$$= 1 - \alpha + n^{-1} \sum_{j=0}^{3} \{\overline{H}_{\mathrm{Rao}}^{(j)} - H_{\mathrm{Rao}}^{(j)}\} \Psi_{p+2j}(z^2) + o(n^{-1})$$

$$= 1 - \alpha + n^{-1} \Big[\frac{1}{2}(W_1 - \overline{W}_1)\{\Psi_p(z^2) - \Psi_{p+2}(z^2)\}$$

$$+ \frac{1}{6}(W_2 - \overline{W}_2)\{\Psi_p(z^2) - 3\Psi_{p+2}(z^2) + 2\Psi_{p+4}(z^2)\} \Big]$$

$$+ o(n^{-1}) . \tag{5.4.9}$$

Since

$$\Psi_\nu(z^2) - \Psi_{\nu+2}(z^2) = 2\psi_{\nu+2}(z^2) = 2\nu^{-1}z^2\psi_\nu(z^2) , \tag{5.4.10}$$

upon further simplification, one gets

$$P^{\overline{\pi}}\{\theta \in Q_{\mathrm{Rao}}^{(1-\alpha)}(\pi, X)|X\}$$

$$= 1 - \alpha + (np)^{-1}z^2\psi_p(z^2)\Big\{W_1 - \overline{W}_1 + \frac{1}{3}(1 - \frac{2z^2}{p+2})(W_2 - \overline{W}_2)\Big\}$$

$$+ o(n^{-1}) .$$

Exactly as in subsection 5.2.2, Steps 2 and 3 of the shrinkage argument now yield

$$P_\theta\{\theta \in Q_{\mathrm{Rao}}^{(1-\alpha)}(\pi, X)\}$$

$$= 1 - \alpha + (np)^{-1}\frac{z^2\psi_p(z^2)}{\pi(\theta)}\Big\{\Delta^{(1)}(\pi, \theta) + \Big(1 - \frac{2z^2}{p+2}\Big)\Delta^{(2)}(\pi, \theta)\Big\}$$

$$+ o(n^{-1}) , \tag{5.4.11}$$

for all θ, where

$$\Delta^{(1)}(\pi, \theta) = I^{jr}\pi_{jr}(\theta) - \pi(\theta)D_jD_r(I^{jr})$$

$$= 2D_r\{I^{jr}\pi_j(\theta)\} - D_jD_r\{\pi(\theta)I^{jr}\} , \tag{5.4.12}$$

using (4.3.8), and

$$\Delta^{(2)}(\pi, \theta) = D_u\{\pi(\theta)L_{jrs}I^{jr}I^{su}\} . \tag{5.4.13}$$

The right-hand side of (5.4.11) equals $1 - \alpha + o(n^{-1})$ for all α and θ if and only if both $\Delta^{(1)}(\pi, \theta)$ and $\Delta^{(2)}(\pi, \theta)$ vanish identically in θ. Thus one gets the following result, due to Rao and Mukerjee (1995), characterizing matching priors associated with $M_{\mathrm{Rao}}(\theta, X)$.

Theorem 5.4.1. *A prior $\pi(\cdot)$ ensures frequentist validity, with margin of error $o(n^{-1})$, of posterior credible regions for θ given by the inversion of Rao's score statistic $M_{Rao}(\theta, X)$ if and only if it satisfies the partial differential equations*

$$2D_r\{I^{jr}\pi_j(\theta)\} - D_jD_r\{\pi(\theta)I^{jr}\} = 0 , \tag{5.4.14}$$

and

$$D_u\{\pi(\theta)L_{jrs}I^{jr}I^{su}\} = 0 . \tag{5.4.15}$$

♣

We next consider Wald's statistic. In the same way as (5.4.4) and (5.4.5), let $Q_{\text{Wald}}^{(1-\alpha)}(\pi, X)$ denote a posterior credible region for θ, with credibility level $1 - \alpha + o(n^{-1})$, as given by the inversion of $M_{\text{Wald}}(\theta, X)$. The approach outlined above yields

$$
\begin{aligned}
&P_\theta\{\theta \in Q_{\text{Wald}}^{(1-\alpha)}(\pi, X)\} \\
&= 1 - \alpha + (np)^{-1}\frac{z^2\psi_p(z^2)}{\pi(\theta)}\left\{\Delta^{(1)}(\pi, \theta) + \left(1 + \frac{z^2}{p+2}\right)\Delta^{(2)}(\pi, \theta)\right\} \\
&\quad + o(n^{-1}) ,
\end{aligned}
\tag{5.4.16}
$$

for all θ, where $\Delta^{(1)}(\pi, \theta)$ and $\Delta^{(2)}(\pi, \theta)$ are given by (5.4.12) and (5.4.13). The truth of (5.4.16) also follows from another derivation in the next section in a somewhat different context. Observe that (5.4.16) is very similar to (5.4.11) and yields the same matching conditions as in Theorem 5.4.1. Thus, as noted in Rao and Mukerjee (1995), the classes of matching priors based on Rao's score and Wald's statistics are identical. Incidentally, Lee (1989) also studied the matching problem associated with Wald's statistic.

Equations (5.4.14) and (5.4.15) add up to the matching condition (5.2.11) for the LR statistic. Therefore, any matching prior arising from Rao's score or Wald's statistic enjoys the same property for the LR statistic as well. The converse, however, is not generally true. This will be clear later in this section when Example 4.3.7 is revisited.

In particular, if $p = 1$ then I is a scalar and, using the regularity condition (2.5.4), the matching conditions (5.4.14) and (5.4.15) reduce to

$$I^{-1}\left(\frac{d\pi(\theta)}{d\theta}\right) - \pi(\theta)(L_{1,11} + L_{111})I^{-2} = \text{constant} , \tag{5.4.17}$$

and

$$\pi(\theta)L_{111}I^{-2} = \text{constant} , \tag{5.4.18}$$

respectively. By (2.5.4), these conditions are met by Jeffreys' prior $\pi(\theta) \propto I^{1/2}$ if and only if

$$I^{-3/2}L_{1,11} = \text{constant} \quad \text{and} \quad I^{-3/2}L_{111} = \text{constant} . \tag{5.4.19}$$

In view of the regularity condition (2.5.5), it is again clear that (5.4.19) entails the corresponding condition (5.2.14) for the LR statistic.

One can readily check that (5.4.19) holds for the one-parameter location or scale models and also for the model in Example 4.3.1. Thus, in these situations, Jeffreys' prior enjoys the matching property for Rao's score and Wald's statistics. On the other hand, in Example 4.3.2 concerning a bivariate normal model with unknown correlation coefficient, not only (5.4.19) fails to hold but also no solution to the matching conditions (5.4.17) and (5.4.18) is available.

Turning to the matching conditions (5.4.14) and (5.4.15) appropriate for general θ, we now revisit the other examples of Section 4.3. The priors listed in (5.2.16), that were seen there to be both HPD and LR matching in Examples 4.3.3, 4.3.4, 4.3.5 and 4.3.7 respectively, satisfy (5.4.14) and (5.4.15) as well in the respective examples. On the other hand, in Example 4.3.6, it is difficult to find any reasonable prior that meets both (5.4.14) and (5.4.15).

We now reconsider Example 4.3.7 in some detail. In the setup of this example, any prior of the form $\pi(\theta) \propto (\theta_1^{r_1} \theta_2^{r_2})^{-1}$, where r_1 and r_2 are real numbers, satisfies (5.4.14) and (5.4.15) if and only if

$$7 - 3r_1 - 4r_2 = 0 \quad \text{and} \quad 4 - r_1 - r_2 - r_1^2 - r_2^2 = 0 \, ,$$

i.e., if and only if (r_1, r_2) equals either (1,1) or (13/25, 34/25). Both these choices of (r_1, r_2) meet the corresponding condition (5.2.17) for the LR statistic. Since there are many other solutions to (5.2.17) as well, this example vindicates our earlier assertion that LR matching priors do not necessarily enjoy the same property for Rao's score or Wald's statistics. If one recalls Example 5.2.1, which is in continuation of Example 4.3.7, it also becomes clear that (1,1) is the unique choice of (r_1, r_2) that satisfies (5.4.14) and (5.4.15) in addition to the HPD and LR matching conditions. Thus the matching conditions (5.4.14) and (5.4.15) can serve as useful supplements to those obtained earlier and facilitate further narrowing down the choice of matching priors.

5.5 Perturbed ellipsoidal and HPD regions

The techniques for characterizing matching priors, discussed so far in this monograph, are useful in addressing a related problem. This concerns finding, for a given prior, credible regions with both Bayesian and frequentist validity up to the order of approximation $o(n^{-1})$. Severini (1993) considered this problem for scalar θ and Ghosh and Mukerjee (1995b) investigated it for general, possibly multidimensional, θ. In this section, we follow the latter authors to explore how, for a given prior, simple perturbations of the ellipsoidal or HPD regions can attain the aforesaid objective. Sweeting (1999, 2001) developed an alternative approach to the problem considered here, via the use of directed likelihood, in the case of scalar θ. We also refer, in this context, to Severini (1994) who studied the closeness of certain frequentist intervals for scalar θ to being Bayesian.

5.5.1 Perturbed ellipsoidal region

Suppose interest lies in the entire parameter vector θ. It was noted in the last section that the inversion of Wald's statistic $M_{\text{Wald}}(\theta, X)$ yields ellipsoidal posterior credible regions. Given a prior $\pi(\cdot)$, by (5.4.3), any such region is of the form

$$\{\theta : n(\theta - \widehat{\theta})^T C(\theta - \widehat{\theta}) \leq k\}$$

where k, which may depend on the prior and the data but not on θ, is dictated by the desired posterior credibility level. Evidently, an ellipsoidal region of this kind is centered at $\widehat{\theta}$. We will examine how, for a given prior $\pi(\cdot)$, a perturbation of the center can yield a credible region with both Bayesian and frequentist validity up to the order of approximation $o(n^{-1})$. To that effect, we write

$$M_{\text{pert}}(\theta, X) = n\{\theta - \widehat{\theta} - n^{-1}m(\widehat{\theta})\}^T C\{\theta - \widehat{\theta} - n^{-1}m(\widehat{\theta})\}, \qquad (5.5.1)$$

and consider a perturbed ellipsoidal region of the form

$$Q_{\text{pert}}^{(1-\alpha)}(\pi, X) = \{\theta : M_{\text{pert}}(\theta, X) \leq k_{1-\alpha}(\pi, X, m)\}, \qquad (5.5.2)$$

where $m(\theta) = (m_1(\theta), \ldots, m_p(\theta))^T$ and $k_{1-\alpha}(\pi, X, m)$ have to be so chosen that the region has both posterior and frequentist coverage probability $1 - \alpha + o(n^{-1})$. It may be made explicit that, for each j $(1 \leq j \leq p)$, $m_j(\cdot)$ is a smooth function of θ with functional form possibly dependent on $\pi(\cdot)$ and α but not on n. Also, $k_{1-\alpha}(\pi, X, m)$ may depend on $\pi(\cdot)$, $X = (X_1, \ldots, X_n)^T$ and $m(\cdot)$ in addition to α, but not on θ.

The explicit determination of (5.5.2) warrants explicit expressions for $m(\cdot)$ and $k_{1-\alpha}(\pi, X, m)$. This is done in two stages. In the first stage, consideration of the posterior coverage enables us to find $k_{1-\alpha}(\pi, X, m)$ explicitly for any given $m(\cdot)$. Then, in the second stage, consideration of the frequentist coverage determines the appropriate choice of $m(\cdot)$.

Let $h = n^{1/2}(\theta - \widehat{\theta})$ and $\widehat{m} = (\widehat{m}_1, \ldots, \widehat{m}_p)^T = m(\widehat{\theta})$. By (5.5.1),

$$M_{\text{pert}}(\theta, X) = (h - n^{-1/2}\widehat{m})^T C(h - n^{-1/2}\widehat{m})$$
$$= h^T Ch - 2n^{-1/2}\widehat{m}^T Ch + n^{-1}\widehat{m}^T C\widehat{m}.$$

Hence the same approach as in the proofs of Lemmas 4.2.1 and 5.2.1 shows that the approximate posterior characteristic function of $M_{\text{pert}}(\theta, X)$, under a prior $\pi(\cdot)$, is given by

$$E^\pi[\exp\{\xi M_{\text{pert}}(\theta, X)\}|X] = (1 - 2\xi)^{-\frac{1}{2}p}\left\{1 + n^{-1}\sum_{j=0}^{3} H_{\text{pert}}^{(j)}(1 - 2\xi)^{-j}\right\}$$
$$+ o(n^{-1}), \qquad (5.5.3)$$

where

$$H^{(0)}_{\text{pert}} = -\frac{1}{72}(36W_1 + 12W_2 + 3W_3 + W_4) - \frac{1}{2}\widehat{m}^T C\widehat{m} + (\widehat{\pi}_j\widehat{m}_j/\widehat{\pi}),$$

$$H^{(1)}_{\text{pert}} = \frac{1}{2}W_1 + \frac{1}{2}\widehat{m}^T C\widehat{m} - (\widehat{\pi}_j\widehat{m}_j/\widehat{\pi}) + \frac{1}{2}a_{jrs}\widehat{m}_j c^{rs},$$

$$H^{(2)}_{\text{pert}} = \frac{1}{24}(4W_2 + W_3) - \frac{1}{2}a_{jrs}\widehat{m}_j c^{rs}, \quad H^{(3)}_{\text{pert}} = \frac{1}{72}W_4,$$

with $W_1, \ldots, W_4$ as shown in (2.2.18). Observe that (5.5.3) has precisely the same form as (5.4.6). Hence, analogously to (5.4.7) and using the same notation as there,

$$k_{1-\alpha}(\pi, X, m) = z^2 - n^{-1}\{\psi_p(z^2)\}^{-1}\left\{\sum_{j=0}^{3} H^{(j)}_{\text{pert}}\Psi_{p+2j}(z^2)\right\} \qquad (5.5.4)$$

ensures that the perturbed ellipsoidal region (5.5.2) has posterior coverage probability $1 - \alpha + o(n^{-1})$.

The shrinkage argument can now be employed as usual to study the frequentist coverage of (5.5.2) with $k_{1-\alpha}(\pi, X, m)$ as in (5.5.4). The details are quite similar to those for Rao's score statistic as discussed in the last section. In particular, one has to execute calculations similar to those in (5.4.9) and employ (5.4.10). Eventually, one gets

$$P_\theta\{\theta \in Q^{(1-\alpha)}_{\text{pert}}(\pi, X)\}$$

$$= 1 - \alpha + (np)^{-1}\frac{z^2\psi_p(z^2)}{\pi(\theta)}$$

$$\times \left[\Delta^{(1)}(\pi, \theta) + \left(1 + \frac{z^2}{p+2}\right)\Delta^{(2)}(\pi, \theta) - 2D_j\{\pi(\theta)m_j(\theta)\}\right]$$

$$+ o(n^{-1}), \qquad (5.5.5)$$

for all θ, where $\Delta^{(1)}(\pi, \theta)$ and $\Delta^{(2)}(\pi, \theta)$ are as in (5.4.12) and (5.4.13). In particular, if $m(\theta) = 0$ identically in θ, then (5.5.1) reduces to Wald's statistic and (5.5.5) reduces to the corresponding equation (5.4.16).

By (5.5.5), the perturbed ellipsoidal region (5.5.2) has frequentist coverage probability $1 - \alpha + o(n^{-1})$ if and only if $m(\theta)$ satisfies the partial differential equation

$$\Delta^{(1)}(\pi, \theta) + \left(1 + \frac{z^2}{p+2}\right)\Delta^{(2)}(\pi, \theta) - 2D_j\{\pi(\theta)m_j(\theta)\} = 0. \qquad (5.5.6)$$

From (5.4.12) and (5.4.13), it is not hard to see that $\Delta^{(1)}(\pi, \theta)$ and $\Delta^{(2)}(\pi, \theta)$ can as well be expressed as

$$\Delta^{(1)}(\pi, \theta) = D_j[D_r\{\pi(\theta)I^{jr}\} - 2\pi(\theta)(D_r I^{jr})],$$

$$\Delta^{(2)}(\pi, \theta) = D_j\{\pi(\theta)L_{rsu}I^{jr}I^{su}\}.$$

Hence (5.5.6) is equivalent to

$$D_j \Big[D_r \{\pi(\theta) I^{jr}\} - 2\pi(\theta)(D_r I^{jr})$$

$$+ \Big(1 + \frac{z^2}{p+2}\Big) \pi(\theta) L_{rsu} I^{jr} I^{su} - 2\pi(\theta) m_j(\theta) \Big] = 0 . \qquad (5.5.7)$$

A particular solution to (5.5.7) for $m(\theta)$ is clearly given by $\overline{m}(\theta)$ where the jth component of $\overline{m}(\theta)$, $1 \le j \le p$, is

$$\overline{m}_j(\theta) = \frac{1}{2} \{\pi(\theta)\}^{-1} D_r \{\pi(\theta) I^{jr}\} - D_r I^{jr} + \frac{1}{2}\Big(1 + \frac{z^2}{p+2}\Big) L_{rsu} I^{jr} I^{su} . \quad (5.5.8)$$

Thus if one obtains $m(\theta)$ by solving (5.5.7) and, with $m(\theta)$ so determined, finds $k_{1-\alpha}(\pi, X, m)$ via (5.5.4) then the perturbed ellipsoidal region (5.5.2) has both posterior and frequentist coverage probability $1 - \alpha + o(n^{-1})$. In general, (5.5.7) has infinitely many solutions. In order to make a choice from amongst rival solutions to (5.5.7), one can consider a principle of minimal perturbation which is sensible from a Bayesian point of view. Thus, given $\pi(\cdot)$, one should first check if $m(\theta) = 0$ satisfies (5.5.7). If not, then using $I = I(\theta)$ as a Riemannian metric a solution with a smaller value of

$$G(m, \pi) = \int \{m(\theta)\}^T I(\theta) \{m(\theta)\} \pi(\theta) d\theta \qquad (5.5.9)$$

will be preferred to another with a larger value of the same quantity. For $p > 1$, it is difficult to characterize all the solutions to (5.5.7) and one may consider using a solution which at least satisfies $G(m, \pi) < \infty$.

Example 5.5.1. Consider the one-parameter scale model (2.5.11). Then $p = 1$, $\theta = \theta_1$, and both I and $m(\cdot)$ are scalars. Furthermore, $I = r_1/\theta^2$ and $L_{111} = r_2/\theta^3$, where $r_1 (> 0)$ and r_2 are constants free from θ. Let $\pi(\theta) = e^{-\theta}$, $\theta > 0$. Then (5.5.7) reduces to

$$\frac{d}{d\theta}\Big[\theta e^{-\theta}\Big\{\Big(1 + \frac{1}{3}z^2\Big) r_1^{-2} r_2 - r_1^{-1}(\theta + 2)\Big\} - 2e^{-\theta} m(\theta)\Big] = 0 ,$$

with a general solution

$$m(\theta) = r_0 e^\theta + \frac{1}{2}\theta\Big\{\Big(1 + \frac{1}{3}z^2\Big) r_1^{-2} r_2 - r_1^{-1}(\theta + 2)\Big\} , \qquad (5.5.10)$$

where r_0 is any constant free from θ. By (5.5.9), a solution of the form (5.5.10) has a finite $G(m, \pi)$ if and only if $r_0 = 0$. Thus the principle of minimal perturbation suggests taking $r_0 = 0$. With this choice of r_0, it is easily seen that (5.5.10) equals the particular solution given by (5.5.8).   ♣

Example 5.5.2. Consider the p-variate normal model with unknown mean vector $\theta = (\theta_1, \ldots, \theta_p)^T$ ($\in \mathcal{R}^p$) and known covariance matrix $\mathcal{I}_p$, where $\mathcal{I}_p$ is the identity matrix of order p. Then $I = \mathcal{I}_p$ and $L_{rsu} = 0$ ($1 \le r, s, u \le p$). Let

$\pi(\cdot)$ be the p-variate normal prior with null mean vector and known, positive definite covariance matrix Ω. One can verify that $m(\theta) = 0$ does not satisfy (5.5.7) and that the particular solution (5.5.8) reduces to $\overline{m}(\theta) = -\frac{1}{2}\Omega^{-1}\theta$. From (5.5.9), it is satisfying to note that $G(\overline{m}, \pi) < \infty$. ♣

Example 5.5.3. Consider the exponential regression model of Example 2.6.5 which was revisited in Example 4.3.4. The I_{su} and L_{rsu} $(r, s, u = 1, 2)$ are as shown in Example 4.3.4. Let $\pi(\theta) = \phi(\theta_1)e^{-\theta_2}(\theta_1 \in \mathcal{R}^1, \theta_2 > 0)$ where, as usual, $\phi(\cdot)$ is the standard normal density. It can be seen that $m(\theta) = 0$ does not satisfy (5.5.7) and that the particular solution (5.5.8) simplifies to

$$\overline{m}_1(\theta) = \frac{1}{2}\gamma_2^{-2}\left\{\left(1 + \frac{1}{4}z^2\right)\gamma_3 - \gamma_2\theta_1\right\},$$
$$\overline{m}_2(\theta) = \frac{1}{2t}\theta_2\left\{5\left(1 + \frac{1}{4}z^2\right) - (2 + \theta_2)\right\},$$

where $\gamma_u = \sum_{j=1}^t z_j^u$ $(u = 2, 3)$. Again, to our satisfaction, $G(\overline{m}, \pi) < \infty$. ♣

Ghosh and Mukerjee (1995b) discussed the extension of the results of this subsection to the case where interest lies in a subvector of θ. The reader may see the original paper for details.

5.5.2 Perturbed HPD region

The ideas of the previous subsection are readily applicable when one considers an HPD region instead of an ellipsoidal region. From (4.2.1), recall that an HPD region is given by the inversion of $M(\theta, \pi, X)$ as defined in (4.2.2). An expansion for $M(\theta, \pi, X)$ was given in (4.2.5). The leading term of this expansion, namely,

$$h^T C h = n(\theta - \widehat{\theta})^T C(\theta - \widehat{\theta})$$

can be perturbed as in (5.5.1). Analogously to (5.5.2), this leads to the consideration of a perturbed HPD region of the form

$$\{\theta : M_{\text{pert}}(\theta, \pi, X) \leq k_{1-\alpha}(\pi, X, m)\},$$

where

$$M_{\text{pert}}(\theta, \pi, X) = M(\theta, \pi, X) - n(\theta - \widehat{\theta})^T C(\theta - \widehat{\theta})$$
$$+ n\{\theta - \widehat{\theta} - n^{-1}m(\widehat{\theta})\}^T C\{\theta - \widehat{\theta} - n^{-1}m(\widehat{\theta})\},$$

with $m(\theta) = (m_1(\theta), \ldots, m_p(\theta))^T$.

An algebra quite similar to that in the last subsection shows that if $m(\theta)$ is obtained by solving the partial differential equation

$$D_j[\pi(\theta)L_{rsu}I^{jr}I^{su} - D_r\{\pi(\theta)I^{jr}\} - 2\pi(\theta)m_j(\theta)] = 0, \qquad (5.5.11)$$

and thereafter $k_{1-\alpha}(\pi, X, m)$ is chosen as

$$
k_{1-\alpha}(\pi, X, m) = z^2 \left[1 + (np)^{-1} \left\{ \frac{1}{12} \left(W_3 + \frac{1}{3} W_4 \right) + \hat{m}^T C \hat{m} - a_{jrs} \hat{m}_j c^{rs} \right\} \right] - 2n^{-1} (\hat{\pi}_j \hat{m}_j / \hat{\pi}),
$$

where $\hat{m} = (\hat{m}_1, \ldots, \hat{m}_p)^T = m(\hat{\theta})$, z^2 is as in (5.4.7) or (5.5.4), and W_3 and W_4 are as shown in (2.2.18), then the perturbed HPD region considered above has both posterior and frequentist coverage probability $1 - \alpha + o(n^{-1})$. More details of the derivation can be found in Ghosh and Mukerjee (1995b) who reported (5.5.11) in another equivalent form.

Clearly, (5.5.11) has infinitely many solutions and a particular solution is given by $\overline{m}(\theta) = (\overline{m}_1(\theta), \ldots, \overline{m}_p(\theta))^T$, where

$$
\overline{m}_j(\theta) = \frac{1}{2} \left[L_{rsu} I^{jr} I^{su} - \{\pi(\theta)\}^{-1} D_r \{\pi(\theta) I^{jr}\} \right], \quad 1 \le j \le p. \qquad (5.5.12)
$$

As before, the principle of minimal perturbation, via consideration of $G(m, \pi)$ defined in (5.5.9), can facilitate a choice from amongst rival solutions to (5.5.11). In each of Examples 5.5.1, 5.5.2 and 5.5.3, the solution $\overline{m}(\theta)$ given by (5.5.12) satisfies $G(\overline{m}, \pi) < \infty$. Furthermore, in Example 5.5.1, it is the only solution with this property.

6

Matching Priors for Prediction

6.1 Introduction

In the preceding chapters, we considered probability matching priors for estimation. The object of interest was a parameter, either one-dimensional or multidimensional, and priors ensuring approximate frequentist validity of the associated posterior credible regions were studied. Evidently, the solutions for these matching priors depend on the specification of the interest parameter. For instance, in Example 2.5.7 concerning the Student's t-model, it was seen that a unique second order matching prior exists when the location parameter θ_1 is of interest whereas no such prior is available when interest lies in the shape parameter θ_2.

A natural alternative approach is to match asymptotically the coverage probability of a Bayesian credible set for a future observation with the corresponding frequentist probability. This is particularly attractive when the main problem is prediction, and not estimation, and thus there is no particular reason to treat a certain parameter as the parameter of interest in preference to others. Our focus in this chapter is on matching priors for prediction.

In Section 6.2, based on the work of Datta, Mukerjee, Ghosh and Sweeting (2000), we characterize priors ensuring approximate frequentist validity of Bayesian predictive regions constructed via posterior quantiles or highest posterior predictive density. Drawing on the work of Datta and Mukerjee (2003), in Section 6.3 we extend the results of Section 6.2 to the regression problem. In this setup each observation involves a dependent variable and an independent variable (auxiliary variable), both possibly vector-valued. We often have knowledge of both the variables in the past observations and also of the independent variable in a new observation. The prediction problem involves the dependent variable in the new observation. We emphasize that unlike in Section 6.3, we do not have any auxiliary variable in Section 6.2. The examples in Sections 6.2 and 6.3 are borrowed from the two papers mentioned above. Finally, we conclude this chapter and also the monograph in Section 6.4 with some brief remarks on areas where more work is needed.

Results of Sections 6.2 and 6.3 provide some theoretical insight into the relationship between Bayesian and frequentist approaches to predictive inference. For a study of reference priors for prediction using an information theoretic approach we refer to Kuboki (1998); see also Sweeting, Datta and Ghosh (2003) for further related results. General discussion on the problem of prediction is available in the books by Aitchison and Dunsmore (1975) and Geisser (1993); see also Barndorff-Nielsen and Cox (1996), Vidoni (1998) and Corcuera and Giummole (1999), among others, for more recent results and further references.

6.2 Matching priors for prediction: no auxiliary variable

In this section, we characterize priors ensuring approximate frequentist validity of Bayesian predictive regions for a future observation. This is done in the absence of any auxiliary variable. We begin by obtaining in subsection 6.2.1 an expansion for the posterior predictive density of the future observation. In the spirit of Chapters 2 and 4, this expansion is used to characterize matching priors associated with two types of Bayesian predictive sets, namely, those given by (a) posterior quantiles, and (b) highest posterior predictive density regions. These problems are addressed in subsections 6.2.2 and 6.2.3 respectively. In particular, the results presented in subsection 6.2.3 are quite different from their counterparts in the context of estimation as reported in Chapter 4.

In subsection 6.2.4, we consider models indexed by a scalar parameter and explore how, for a given prior, one can construct prediction intervals for which Bayesian and frequentist coverage probabilities match approximately. This provides a predictive analog of the developments in Section 5.5 but the approach here is entirely different.

6.2.1 Preliminaries: expansion for the predictive density

Let $X_1, X_2, \ldots$ be a sequence of i.i.d. possibly vector-valued random variables with a common density $f(x; \theta)$. We consider Bayesian prediction of X_{n+1}, with approximate frequentist validity, based on $X = (X_1, \ldots, X_n)^T$, using a prior density $\pi(\cdot)$. Thus X_{n+1} is regarded as a future observation whereas $X_1, \ldots, X_n$ are the past, and hence available, ones. The setup and assumptions are again as in Section 2.2. Unless otherwise stated, the notation is also as described there.

Let $\widetilde{\pi}(x_{n+1}|X)$ denote the posterior predictive density of X_{n+1} given X under the prior $\pi(\cdot)$. As in Section 2.2, let $h = (h_1, \ldots, h_p)^T = n^{1/2}(\theta - \widehat{\theta})$, where $\widehat{\theta}$ is the maximum likelihood estimator of θ based on X, and write $\pi(\theta|X)$ and $\pi^*(h|X)$ respectively for the posterior densities of θ and h, given X, under $\pi(\cdot)$. Then

$$\tilde{\pi}(x_{n+1}|X) = \int f(x_{n+1};\theta)\pi(\theta|X)\mathrm{d}\theta$$

$$= \int f(x_{n+1};\hat{\theta} + n^{-1/2}h)\pi^*(h|X)\mathrm{d}h . \qquad (6.2.1)$$

We now derive an expansion for $\tilde{\pi}(x_{n+1}|X)$ up to the order of approximation $o(n^{-1})$.

Let $f_j(x;\theta) = D_j f(x;\theta)$ and $f_{jk}(x;\theta) = D_j D_k f(x;\theta)$. Using Taylor's expansion for $f(x_{n+1};\hat{\theta} + n^{-1/2}h)$ around $\hat{\theta}$ we get

$$f(x_{n+1};\hat{\theta} + n^{-1/2}h) = f(x_{n+1};\hat{\theta}) + n^{-1/2}f_j(x_{n+1};\hat{\theta})h_j$$

$$+ \frac{1}{2}n^{-1}f_{jk}(x_{n+1};\hat{\theta})h_j h_k + o(n^{-1}) . \qquad (6.2.2)$$

By (6.2.2) and (2.2.19), we get from (6.2.1) that

$$\tilde{\pi}(x_{n+1}|X) = f(x_{n+1};\hat{\theta}) + n^{-1}f_j(x_{n+1};\hat{\theta})$$

$$\times \int h_j\left\{R_1(h) + \frac{1}{6}R_3(h)\right\}\phi_p(h;C^{-1})\mathrm{d}h$$

$$+ \frac{1}{2}n^{-1}f_{jk}(x_{n+1};\hat{\theta})\int h_j h_k \phi_p(h;C^{-1})\mathrm{d}h$$

$$+ o(n^{-1}) , \qquad (6.2.3)$$

where $R_1(h)$ and $R_3(h)$ are as given by (2.2.12). Clearly,

$$\int h_j h_k \phi_p(h;C^{-1})\mathrm{d}h = c^{jk} ,$$

and, recalling (2.2.12),

$$\int h_j\left\{R_1(h) + \frac{1}{6}R_3(h)\right\}\phi_p(h;C^{-1})\mathrm{d}h = c^{jr}\frac{\tilde{\pi}_r}{\tilde{\pi}} + \frac{1}{2}c^{jk}c^{rs}a_{krs} .$$

From (6.2.3), it now follows that

$$\tilde{\pi}(x_{n+1}|X) = f(x_{n+1};\hat{\theta}) + n^{-1}f_j(x_{n+1};\hat{\theta})c^{jr}\frac{\tilde{\pi}_r}{\tilde{\pi}} + n^{-1}b(x_{n+1},X) + o(n^{-1}) ,$$

$$(6.2.4)$$

where

$$b(x_{n+1},X) = \frac{1}{2}\left[f_{jk}(x_{n+1};\hat{\theta}) + f_j(x_{n+1};\hat{\theta})c^{rs}a_{krs}\right]c^{jk} . \qquad (6.2.5)$$

The expansion (6.2.4) is in agreement with the findings in Komaki (1996) and Corcuera and Giummole (1999) in a different context.

Since the leading term in the right-hand side of (6.2.4) is usually non-normal, the results for prediction differ from those arising in the context of estimation. We show in the next subsection that for prediction of a scalar valued random variable X_{n+1} based on quantiles, even in the scalar parameter case, Jeffreys' prior does not automatically emerge as a solution to a differential equation but, as Theorem 6.2.1 below reveals, a new approach is needed in examining its role. The explicit characterizations obtained later in this section help in understanding how far Jeffreys' prior can yield asymptotically valid frequentist inference for the problem of prediction and demonstrate that it works only in some but not all situations.

6.2.2 Frequentist validity of posterior quantiles

In this subsection, we consider the case where the X_i, $i \geq 1$, are scalar-valued. Then the posterior quantiles of X_{n+1}, given $X = (X_1, \ldots, X_n)^T$ are meaningful. We assume that the set $\{u : f(u; \theta) > 0\}$ is an interval, possibly unbounded, on the real line. Then for each θ and α $(0 < \alpha < 1)$, there exists a unique $q(\theta, \alpha)$ such that

$$\int_{q(\theta,\alpha)}^{\infty} f(u; \theta)du = \alpha . \tag{6.2.6}$$

Define

$$\mu_j(\theta, \alpha) = \int_{q(\theta,\alpha)}^{\infty} f_j(u; \theta)du . \tag{6.2.7}$$

Let $Q(\pi, X, \alpha)$ denote the $(1-\alpha)$th posterior quantile of X_{n+1} under the prior $\pi(\cdot)$. Then, from (6.2.4) and (6.2.6), it follows that

$$Q(\pi, X, \alpha) = q(\hat{\theta}, \alpha) + O(n^{-1}) . \tag{6.2.8}$$

We now proceed to characterize priors ensuring approximate frequentist validity of the posterior quantiles of X_{n+1}. The shrinkage argument, as given in Steps 1–3 of Section 1.2, is again helpful. We take an auxiliary prior $\overline{\pi}(\cdot)$ which satisfies the conditions in Bickel and Ghosh (1990) and is as described in Section 2.4. As usual, write $P^\pi\{.|X\}$ and $P^{\overline{\pi}}\{.|X\}$ for the posterior probability measures under $\pi(\cdot)$ and $\overline{\pi}(\cdot)$ respectively. Since $b(., X)$ as shown in (6.2.5) does not depend on $\pi(\cdot)$, from (6.2.4), (6.2.7) and (6.2.8) one gets

$$\begin{aligned}
&P^{\overline{\pi}}\{X_{n+1} > Q(\pi, X, \alpha)|X\} \\
&= P^\pi\{X_{n+1} > Q(\pi, X, \alpha)|X\} \\
&\quad + n^{-1}c^{jr}\left(\frac{\overline{\pi}_r}{\overline{\pi}} - \frac{\hat{\pi}_r}{\hat{\pi}}\right)\int_{Q(\pi,X,\alpha)}^{\infty} f_j(u; \hat{\theta})du + o(n^{-1}) \\
&= \alpha + n^{-1}c^{jr}\left(\frac{\overline{\pi}_r}{\overline{\pi}} - \frac{\hat{\pi}_r}{\hat{\pi}}\right)\mu_j(\hat{\theta}; \alpha) + o(n^{-1}) , \tag{6.2.9}
\end{aligned}$$

in Step 1. Considering Step 2 of the shrinkage argument, in view of (2.2.1)–(2.2.3), it is immediate from (6.2.9) that

$$E_\theta P^{\overline{\pi}}\{X_{n+1} > Q(\pi, X, \alpha)|X\} = \alpha + n^{-1}I^{jr}\mu_j(\theta, \alpha)\left\{\frac{\overline{\pi}_r(\theta)}{\overline{\pi}(\theta)} - \frac{\pi_r(\theta)}{\pi(\theta)}\right\}$$
$$+ o(n^{-1}), \tag{6.2.10}$$

for θ in the interior of the support of $\overline{\pi}(\cdot)$. By (6.2.10), proceeding as in Section 2.4, Step 3 now yields

$$P_\theta\{X_{n+1} > Q(\pi, X, \alpha)\} = \alpha - \frac{1}{n\pi(\theta)}D_r\{I^{jr}\mu_j(\theta, \alpha)\pi(\theta)\} + o(n^{-1}), \tag{6.2.11}$$

for all θ. The right-hand side of (6.2.11) equals $\alpha + o(n^{-1})$ if and only if

$$D_r\{I^{jr}\mu_j(\theta, \alpha)\pi(\theta)\} = 0. \tag{6.2.12}$$

A prior $\pi(\cdot)$, satisfying the partial differential equation (6.2.12) for every α, is a matching prior in the sense of ensuring frequentist validity, up to $o(n^{-1})$, of the posterior quantiles of X_{n+1}.

In contrast with the partial differential equations (2.4.11) and (2.4.12) characterizing matching priors for estimation, the corresponding equation (6.2.12) for prediction is not always free from α. As noted in subsection 2.5.1, for scalar parameter models a first order matching prior for estimation is uniquely given by Jeffreys' prior. We now examine the role of Jeffreys' prior in the context of prediction.

Theorem 6.2.1. *For scalar θ, if there exists a prior satisfying (6.2.12) for every α then it must be Jeffreys' prior.*

Proof. Differentiation of both sides of (6.2.6) with respect to α yields

$$-f(q(\theta, \alpha); \theta)\frac{\partial q(\theta, \alpha)}{\partial \alpha} = 1. \tag{6.2.13}$$

Suppose there exists a prior, say $\pi_0(\theta)$, satisfying (6.2.12) for every α. Then, with $I \equiv I(\theta)$,

$$\mu_1(\theta, \alpha) = \psi(\alpha)\frac{I(\theta)}{\pi_0(\theta)}, \tag{6.2.14}$$

where $\psi(\alpha)$ does not involve the scalar parameter θ. Also, from (6.2.7),

$$\mu_1(\theta, \alpha) = \int_{q(\theta, \alpha)}^{\infty} f_\theta(x; \theta)dx, \tag{6.2.15}$$

where $f_\theta(x; \theta) = \partial f(x; \theta)/\partial\theta$. Transforming $x = q(\theta, \beta)$ in the right-hand side of (6.2.15) and using (6.2.13), we get

$$\mu_1(\theta, \alpha) = \int_0^\alpha \frac{f_\theta(q(\theta, \beta); \theta)}{f(q(\theta, \beta); \theta)}d\beta. \tag{6.2.16}$$

By (6.2.14) and (6.2.16),

$$\int_0^\alpha \frac{f_\theta(q(\theta,\beta);\theta)}{f(q(\theta,\beta);\theta)} d\beta = \psi(\alpha)\frac{I(\theta)}{\pi_0(\theta)} .$$

Differentiating both sides of the last equation with respect to α,

$$\frac{f_\theta(q(\theta,\alpha);\theta)}{f(q(\theta,\alpha);\theta)} = \frac{d}{d\alpha}\psi(\alpha)\frac{I(\theta)}{\pi_0(\theta)} . \qquad (6.2.17)$$

Now observe that, analogously to (6.2.16),

$$I(\theta) = \int_{-\infty}^\infty \frac{\{f_\theta(x;\theta)\}^2}{f(x;\theta)} dx = \int_0^1 \left\{ \frac{f_\theta(q(\theta,\beta);\theta)}{f(q(\theta,\beta);\theta)} \right\}^2 d\beta . \qquad (6.2.18)$$

Use of (6.2.17) in (6.2.18) shows that $I(\theta) \propto \{I(\theta)/\pi_0(\theta)\}^2$, that is, $\pi_0(\theta) \propto \{I(\theta)\}^{1/2}$, which proves the result. ♣

Example 6.2.1. Consider the one-parameter scale model of (2.5.11) as given by $f(x;\theta) = \theta^{-1}f^*(x/\theta)$, where $\theta > 0$ and $f^*(\cdot)$ is a density. Let k_α be such that $\int_{k_\alpha}^\infty f^*(u)du = \alpha$. Then by (6.2.6) and (6.2.7), $q(\theta,\alpha) = k_\alpha\theta$, $\mu_1(\theta,\alpha) = \theta^{-1}k_\alpha f^*(k_\alpha)$. Also, $I \propto \theta^{-2}$. Hence $\pi(\theta) \propto \theta^{-1}$, which is Jeffreys' prior, emerges as the unique solution to (6.2.12) for every α. In a similar manner, it is not hard to see that the same conclusion about Jeffreys' prior holds also for the one-parameter location model given by (2.5.6). In fact, for the one-parameter location or scale models the matching property of Jeffreys' prior, as noted in this example, is exact. This can be shown proceeding along the line of Theorem 2.5.2 and we refer to Datta, Ghosh and Mukerjee (2000) for details. Additional related exact results are available in Fraser and Reid (2002). ♣

Example 6.2.2. Even outside one-parameter location or scale models, Jeffreys' prior may satisfy (6.2.12) for every α. This happens for the model specified by

$$f(x;\theta) = \frac{\theta(1+\theta)}{(x+\theta)^2} , \quad 0 < x < 1 , \quad \theta > 0 .$$

It can be checked that here

$$q(\theta,\alpha) = \frac{(1-\alpha)\theta}{\alpha+\theta} , \quad \mu_1(\theta,\alpha) = \frac{\alpha(1-\alpha)}{\theta(1+\theta)} , \quad I = \frac{1}{3\theta^2(1+\theta)^2} .$$

Hence $\pi(\theta) \propto \theta^{-1}(1+\theta)^{-1}$ uniquely satisfies (6.2.12) for every α. Clearly, this is Jeffreys' prior. ♣

There are one-parameter models where Jeffreys' prior does not satisfy (6.2.12) and hence, by Theorem 6.2.1, no solution to (6.2.12), valid for every α, is available. The following example serves as an illustration.

Example 6.2.3. Let $f(x; \theta)$ represent the univariate normal model with both mean and variance equal to θ (> 0). Then $I = (2\theta + 1)/(2\theta^2)$, and by (6.2.6) and (6.2.7),

$$q(\theta, \alpha) = \theta + z_\alpha \theta^{1/2} , \quad \mu_1(\theta, \alpha) = \phi(z_\alpha)(2\theta^{1/2} + z_\alpha)/(2\theta) ,$$

where, as usual, $\phi(\cdot)$ is the standard univariate normal density and z_α is the corresponding $(1 - \alpha)$th quantile. Hence it can be seen that for no α Jeffreys' prior satisfies (6.2.12). ♣

Theorem 6.2.1 does not hold for vector-valued θ. As illustrated by the next example, there it is possible that Jeffreys' prior does not satisfy (6.2.12) but another solution to (6.2.12), valid for every α, is available.

Example 6.2.4. Consider the location-scale model of (2.5.16) as given by $f(x; \theta) = \theta_2^{-1} f^*((x - \theta_1)/\theta_2)$, where $\theta_1 \in \mathcal{R}^1$, $\theta_2 > 0$, and $f^*(\cdot)$ is a density with support $\mathcal{R}^1$. Let k_α be such that $\int_{k_\alpha}^\infty f^*(u)du = \alpha$. Then by (6.2.6) and (6.2.7), $q(\theta, \alpha) = \theta_1 + k_\alpha \theta_2$, $\mu_1(\theta, \alpha) = \theta_2^{-1} f^*(k_\alpha)$, $\mu_2(\theta, \alpha) = \theta_2^{-1} k_\alpha f^*(k_\alpha)$. Also, $I^{jr} = u^{jr} \theta_2^2$ for each j, r, where u^{jr} is free from θ. Hence (6.2.12) reduces to

$$u^{1r} D_r \{\theta_2 \pi(\theta)\} + k_\alpha u^{2r} D_r \{\theta_2 \pi(\theta)\} = 0 . \tag{6.2.19}$$

The condition (6.2.19) holds for every α if and only if $u^{jr} D_r \{\theta_2 \pi(\theta)\} = 0$ ($j = 1, 2$), i.e., if and only if

$$D_r \{\theta_2 \pi(\theta)\} = 0 \ (r = 1 , \ 2) , \tag{6.2.20}$$

invoking the positive definiteness of I and hence that of the matrix $((u^{jr}))$. The prior $\pi(\theta) \propto \theta_2^{-1}$ uniquely satisfies (6.2.20) and hence (6.2.12) for every α. As noted in subsection 2.5.2, this prior also ensures frequentist validity, up to $o(n^{-1})$, of the posterior quantiles of both θ_1 and θ_2. Furthermore, in Chapters 3–5, it was seen to enjoy the matching property in various other senses as well. ♣

Before concluding this subsection, we briefly touch upon the issue of invariance with respect to the parameterization. Let λ be a one-to-one transformation of θ and $\pi^*(\lambda)$ be the transformed version, under the λ−parameterization, of a prior $\pi(\theta)$. It is readily seen that the posterior predictive density of the future observation X_{n+1} under $\pi(\theta)$ and the θ−parameterization is the same as that under $\pi^*(\lambda)$ and the λ−parameterization. Therefore, as in Section 2.7, $\pi(\theta)$ ensures frequentist validity of the posterior quantiles of X_{n+1} up to any order of approximation under the θ−parameterization if and only if $\pi^*(\lambda)$ does so under the λ−parameterization. Thus the matching problem considered in this subsection is invariant of the parameterization adopted.

6.2.3 Frequentist validity of highest posterior predictive density regions

We now turn to the general case where the X_i, $i \geq 1$, are possibly vector-valued. While the posterior quantiles of X_{n+1} are well-defined for scalar X_i, they do not remain so with vector X_i. Even in the latter situation, however, one may consider a highest posterior predictive density (HPPD) region for predicting X_{n+1}. We now explore the conditions under which such prediction has approximate frequentist validity.

Let U be a random variable with density $f(\cdot;\theta)$. We assume that, for each θ, $f(U;\theta)$ has a density which is positive over an interval on the real line, and zero elsewhere. Then for each θ and α ($0 < \alpha < 1$), there exists a unique $m(\theta,\alpha)$ such that

$$\int_A f(u;\theta)du = 1 - \alpha , \qquad (6.2.21)$$

where $A \equiv A(\theta,\alpha) = \{u : f(u;\theta) \geq m(\theta,\alpha)\}$. Define $\xi_j(\theta,\alpha)$ by

$$\xi_j(\theta,\alpha) = \int_A f_j(u;\theta)du . \qquad (6.2.22)$$

By (6.2.4) and (6.2.21), analogously to (6.2.8), the HPPD region for X_{n+1} with posterior coverage $1 - \alpha$ under $\pi(\cdot)$ has the form

$$H(\pi,X,\alpha) = \{u : \tilde\pi(u|X) \geq m(\widehat\theta,\alpha) + n^{-1}\rho\} , \qquad (6.2.23)$$

where $\rho \equiv \rho(\pi,X,\alpha)$ may depend on $\pi(\cdot)$, X and α and is at most of order $O(1)$. The explicit form of ρ is not needed in what follows.

The shrinkage argument again facilitates the study of the frequentist coverage of $H(\pi,X,\alpha)$. The execution of Step 1 of this argument is quite similar to that in the last subsection. By (6.2.4), (6.2.22) and (6.2.23), this step yields

$$P^{\overline\pi}\{X_{n+1} \in H(\pi,X,\alpha)|X\}$$

$$= P^\pi\{X_{n+1} \in H(\pi,X,\alpha)|X\} + n^{-1}c^{jr}\left(\frac{\widehat{\overline\pi}_r}{\widehat{\overline\pi}} - \frac{\widehat\pi_r}{\widehat\pi}\right)\int_{H(\pi,X,\alpha)} f_j(u;\widehat\theta)du$$

$$+ o(n^{-1})$$

$$= 1 - \alpha + n^{-1}c^{jr}\left(\frac{\widehat{\overline\pi}_r}{\widehat{\overline\pi}} - \frac{\widehat\pi_r}{\widehat\pi}\right)\xi_j(\widehat\theta;\alpha) + o(n^{-1}) . \qquad (6.2.24)$$

Along the line of the last subsection, using Steps 2 and 3 of the shrinkage argument, from (6.2.24) one gets

$$P_\theta\{X_{n+1} \in H(\pi,X,\alpha)\} = 1 - \alpha - \frac{1}{n\pi(\theta)}D_r\{I^{jr}\xi_j(\theta,\alpha)\pi(\theta)\} + o(n^{-1}) , \qquad (6.2.25)$$

for all θ. The right-hand side of (6.2.25) equals $1 - \alpha + o(n^{-1})$ if and only if

$$D_r\{I^{jr}\xi_j(\theta,\alpha)\pi(\theta)\} = 0 . \qquad (6.2.26)$$

A prior $\pi(\cdot)$, satisfying the partial differential equation (6.2.26) for every α, is a matching prior in the sense of ensuring frequentist validity, up to $o(n^{-1})$, of the HPPD regions for X_{n+1}. The same argument as in the last subsection shows that such a prior is invariant of the parameterization adopted.

Example 6.2.5. We revisit the bivariate normal model considered in Examples 2.5.3 and 2.9.1, and work under the θ−parameterization as shown in (2.5.27). Then by (6.2.21) and (6.2.22)

$$m(\theta, \alpha) = \frac{\alpha}{2\pi\sqrt{(\theta_2\theta_3)}} , \quad \xi_1(\theta, \alpha) = \xi_4(\theta, \alpha) = \xi_5(\theta, \alpha) = 0 ,$$

$$\theta_2\xi_2(\theta, \alpha) = \theta_3\xi_3(\theta, \alpha) = \frac{1}{2}\alpha \log \alpha ,$$

the π in $m(\theta, \alpha)$ being the usual transcendental number (not to be confused with a prior). Using the above in conjunction with the expressions for the I_{jr} given in Example 2.9.1, one can check that (6.2.26) holds for every α if and only if

$$D_2\{\theta_2\pi(\theta)\} + D_3\{\theta_3\pi(\theta)\} = 0 . \tag{6.2.27}$$

Priors of the form $\pi(\theta) \propto \theta_2^{-r}\theta_3^{r-2}$, where r is any real number, satisfy (6.2.27). The special case $r = 1$ yields a prior as envisaged in Example 2.9.1 in the context of matching alternative coverage probabilities when interest lies in the regression coefficient θ_1. ♣

Example 6.2.6. Consider the multivariate scale model with a density of the form

$$f(x; \theta) = (\theta_1 \ldots \theta_p)^{-1} f^*(x^{(1)}/\theta_1, \ldots, x^{(p)}/\theta_p) ,$$

where $x = (x^{(1)}, \ldots, x^{(p)})^T$ and $\theta_1, \ldots, \theta_p > 0$. Let m_α be such that

$$\int f^*(u_1, \ldots, u_p) du = 1 - \alpha ,$$

where the integral is over $\{u = (u_1, \ldots, u_p)^T : f^*(u_1, \ldots, u_p) \geq m_\alpha\}$. Then by (6.2.21) and (6.2.22)

$$m(\theta, \alpha) = m_\alpha/(\theta_1 \ldots \theta_p) , \quad \xi_j(\theta, \alpha) = \omega_j(\alpha)/\theta_j ,$$

with $\omega_j(\alpha)$ free from θ. Also, $I^{jr} \propto \theta_j\theta_r$ for each j, r. Hence it can be seen that Jeffreys' prior $\pi(\theta) \propto (\theta_1 \ldots \theta_p)^{-1}$ satisfies (6.2.26) for every α. Similarly, for the multivariate location model $f(x; \theta) = f^*(x^{(1)} - \theta_1, \ldots, x^{(p)} - \theta_p)$, Jeffreys' prior, given by $\pi(\theta) = $ constant, satisfies (6.2.26) for every α. By Theorem 4.3.1, one can check that under both the multivariate location and scale models Jeffreys' prior is also HPD matching for θ. Recall also that under both these models Jeffreys' prior was found to be c.d.f. matching in Example 3.3.6. ♣

In general, there is no guarantee that Jeffreys' prior will always satisfy (6.2.26) for every α. As Example 6.2.7 below demonstrates, even with scalar θ and scalar X_i, $i \geq 1$, it is possible that Jeffreys' prior does not satisfy (6.2.26) but another solution to (6.2.26), valid for every α, is available.

Example 6.2.7. We revisit Example 6.2.3. By (6.2.21) and (6.2.22), writing $z^* = z_{\frac{1}{2}\alpha}$,

$$m(\theta, \alpha) = \theta^{-1/2}\phi(z^*), \quad \xi_1(\theta, \alpha) = -\theta^{-1}z^*\phi(z^*).$$

Hence the unique prior, satisfying (6.2.26) for every α, is $\pi(\theta) \propto (2\theta + 1)/\theta$. This is different from Jeffreys' prior and, following Theorem 4.3.1, also from HPD matching priors for θ.   ♣

We now indicate some situations where no prior satisfying (6.2.26) for every α is available. Consider the case of scalar θ. Since both $I \equiv I(\theta)$ and $\pi(\theta)$ are positive for all θ, a solution to (6.2.26), valid for every α, is available if and only if $\xi_1(\theta, \alpha)$ is of the form

$$\xi_1(\theta, \alpha) = Q(\theta)R(\alpha), \quad (6.2.28)$$

where $Q(\theta)$ does not involve α, $R(\alpha)$ does not involve θ, and $Q(\theta)$ is positive for all θ. For the truncated exponential model $f(x; \theta) = k(\theta)\exp(-x/\theta)$, $0 < x < 1$, where $k(\theta) = 1/[\theta\{1 - \exp(-1/\theta)\}]$ and $\theta > 0$, a factorization as in (6.2.28) is not possible. On the other hand, for the bivariate normal model with zero means, unit variances and correlation coefficient θ ($|\theta| < 1$) such a factorization is possible but one cannot have $Q(\theta)$ positive over the entire parameter space. Hence in these two situations no solution to (6.2.26), valid for every α, exists.

One can check that the matching priors reported in Examples 6.2.1, 6.2.2 and 6.2.4–6.2.7 entail the propriety of the posterior predictive densities, with P_θ–probability unity for all θ, whenever n is sufficiently large. Incidentally, recall that no matching prior was available in Example 6.2.3.

In the context of group transformation models, Severini, Mukerjee and Ghosh (2002) obtained some exact results on matching priors for prediction, with application to HPPD regions. These results exploit an earlier one due to Hora and Buehler (1966). We refer to the original sources for details.

6.2.4 Prediction intervals

For the case of scalar θ, we will now consider prediction intervals for a scalar-valued future observation X_{n+1}. It turns out that, for any given prior, it is often possible to choose an interval for which the Bayesian and frequentist coverage probabilities match to the order $o(n^{-1})$.

Fix α ($0 < \alpha < 1$) and let $\gamma \equiv \gamma_\alpha(\theta)$ be any function, with functional form free from n, satisfying $1 - \alpha < \gamma_\alpha(\theta) < 1$ for all θ. Write $\widehat{\gamma} = \gamma_\alpha(\widehat{\theta})$. Then $0 < 1 - \widehat{\gamma} < 2 - \widehat{\gamma} - \alpha < 1$, and

$$P^\pi\{Q(\pi, X, 2 - \widehat{\gamma} - \alpha) < X_{n+1} \le Q(\pi, X, 1 - \widehat{\gamma})|X\} = 1 - \alpha,$$

where, as in subsection 6.2.1, $Q(\pi, X, \eta)$ is the $(1 - \eta)$th posterior quantile of X_{n+1} under the prior $\pi(\cdot)$. An algebra similar to the derivation of (6.2.11) shows that

$$P_\theta\{Q(\pi, X, 2 - \widehat{\gamma} - \alpha) < X_{n+1} \leq Q(\pi, X, 1 - \widehat{\gamma})\}$$

$$= 1 - \alpha - \frac{1}{n\pi(\theta)} \frac{\mathrm{d}}{\mathrm{d}\theta} \left\{ I^{-1}\psi(\gamma, \alpha, \theta)\pi(\theta) \right\} + o(n^{-1}),$$

for all θ, where

$$\psi(\gamma, \alpha, \theta) = \int_{q(\theta, 2-\gamma-\alpha)}^{q(\theta, 1-\gamma)} f_\theta(x; \theta)\mathrm{d}x,$$

with $f_\theta(x; \theta) = \partial f(x; \theta)/\partial\theta$, as before. If we can find a function $\gamma \equiv \gamma_\alpha(\theta)$ for which $\psi(\gamma, \alpha, \theta) = 0$, then for any prior the Bayesian and frequentist coverage probabilities of the prediction interval $(Q(\pi, X, 2 - \widehat{\gamma} - \alpha), Q(\pi, X, 1 - \widehat{\gamma})]$ will agree to $o(n^{-1})$. Sufficient conditions for the existence and uniqueness of such a function are given in the next lemma. Here $F(x; \theta)$ is the common cumulative distribution function of the observations $X_1, X_2, \ldots$.

Lemma 6.2.1. *Suppose that, for each θ, the equation $f_\theta(x; \theta) = 0$ has a unique solution $x(\theta)$, and $f_{\theta x}(x(\theta); \theta) \neq 0$, where $f_{\theta x}(x; \theta) = \frac{\partial^2}{\partial x \partial \theta} f(x; \theta)$. Let*

$$1 - \alpha \geq \alpha_0 \qquad (6.2.29)$$

where

$$\alpha_0 = \sup_\theta [\max\{F(x(\theta); \theta), \ 1 - F(x(\theta); \theta)\}].$$

Then there exists a unique solution $\gamma \equiv \gamma_\alpha(\theta)$ in $(1 - \alpha, 1)$ to the equation

$$\psi(\gamma, \alpha, \theta) = 0. \qquad (6.2.30)$$

Proof. Without loss of generality assume that $f_{\theta x}(x(\theta); \theta) > 0$. Then $f_\theta(x; \theta) < f_\theta(x(\theta); \theta) = 0$ for $x < x(\theta)$ and $f_\theta(x; \theta) > 0$ for $x > x(\theta)$. Since

$$\int_{-\infty}^{\infty} f_\theta(x; \theta)\mathrm{d}x = 0,$$

we have

$$\psi(1 - \alpha, \alpha, \theta) = \int_{-\infty}^{q(\theta, \alpha)} f_\theta(x; \theta)\mathrm{d}x = -\int_{q(\theta, \alpha)}^{\infty} f_\theta(x; \theta)\mathrm{d}x < 0 \quad (6.2.31)$$

provided that $q(\theta, \alpha) \geq x(\theta)$, and

$$\psi(1, \alpha, \theta) = \int_{q(\theta, 1-\alpha)}^{\infty} f_\theta(x; \theta)\mathrm{d}x = -\int_{-\infty}^{q(\theta, 1-\alpha)} f_\theta(x; \theta)\mathrm{d}x > 0 \quad (6.2.32)$$

provided that $q(\theta, 1 - \alpha) \leq x(\theta)$. Also, for $1 - \alpha < \gamma < \gamma^* < 1$,

$$\psi(\gamma^*, \ \alpha, \ \theta) - \psi(\gamma, \ \alpha, \ \theta)$$

$$= \int_{q(\theta, \ 2-\gamma^*-\alpha)}^{q(\theta, \ 1-\gamma^*)} f_\theta(x; \theta) dx - \int_{q(\theta, \ 2-\gamma-\alpha)}^{q(\theta, \ 1-\gamma)} f_\theta(x; \theta) dx$$

$$= \int_{q(\theta, \ 1-\gamma)}^{q(\theta, \ 1-\gamma^*)} f_\theta(x; \theta) dx - \int_{q(\theta, \ 2-\gamma-\alpha)}^{q(\theta, \ 2-\gamma^*-\alpha)} f_\theta(x; \theta) dx$$

$$> 0 \ .$$

The last inequality is justified by the fact that for x in $(q(\theta, \ 1-\gamma), \ q(\theta, \ 1-\gamma^*))$, one has $x > q(\theta, \ 1-\gamma) > q(\theta, \ \alpha) \geq x(\theta)$, implying thereby $f_\theta(x; \theta) > 0$. Similarly, for x in $(q(\theta, \ 2-\gamma-\alpha), \ q(\theta, \ 2-\gamma^*-\alpha))$, one has $x < q(\theta, \ 2-\gamma^*-\alpha) < q(\theta, \ 1-\alpha) \leq x(\theta)$, implying thereby $f_\theta(x; \theta) < 0$. Hence the function $\psi(\gamma, \ \alpha, \ \theta)$ is increasing in γ; so (6.2.31) and (6.2.32) imply that there is a unique function $\gamma_\alpha(\theta)$ with $1 - \alpha < \gamma_\alpha(\theta) < 1$ for which $\psi(\gamma_\alpha(\theta), \ \alpha, \ \theta) = 0$. Finally, the conditions required for (6.2.31) and (6.2.32) are readily seen to be equivalent to (6.2.29), which completes the proof.   ♣

Computation of $\gamma_\alpha(\widehat{\theta})$ can easily be achieved using a Newton iteration. Even though the condition (6.2.29) appears to be restrictive, it is met quite commonly. In fact, as the next example illustrates, even when this condition is not met, there may exist $\gamma_\alpha(\theta)$ in $(1 - \alpha, 1)$ satisfying (6.2.30).

Example 6.2.8. Consider the simple exponential model given by

$$f(x; \theta) = \theta^{-1} \exp(-x/\theta) \ , \ x > 0 \ ,$$

where $\theta > 0$. It can be checked that $x(\theta) = \theta$ and $F(x(\theta); \theta) = 1 - e^{-1}$. Thus $\alpha_0 = 1 - e^{-1}$. For $1 - \alpha = 0.9$ or 0.95, condition (6.2.29) is satisfied. In fact, considering $\psi(\gamma, \ \alpha, \ \theta)$ explicitly, it can be seen that in this example a unique $\gamma_\alpha(\theta)$, as envisaged in (6.2.30), exists for every α in $(0,1)$. This $\gamma_\alpha(\theta)$ does not actually involve θ and is given by the unique solution in $(1 - \alpha, 1)$ for γ to

$$(1 - \gamma) \log(1 - \gamma) - (2 - \gamma - \alpha) \log(2 - \gamma - \alpha) = \ 0 \ .$$

Consider now any prior of the form $\pi(\theta) \propto 1/\theta^r$. Then from the exact posterior predictive density which is easy to work out in the present example, it can be seen that the prediction interval considered in this subsection takes the natural form $(k_1 \overline{X}, \ k_2 \overline{X})$, where $\overline{X}$ is the arithmetic mean of $X_1, \cdots, X_n$, and k_1 and k_2 are constants which involve α and r.   ♣

6.3 Matching priors for predicting a dependent variable in regression models

In this section, we consider Bayesian prediction of a dependent variable given an independent auxiliary variable and past observations on the two variables.

This problem, arising in regression settings, is different from the one discussed in the last section where no distinction between such dependent and independent variables was made. Starting from (6.2.4), we obtain an expansion for the relevant posterior predictive density in subsection 6.3.1. This expansion is used in subsection 6.3.2 to indicate matching conditions, pertaining to the present problem, on the basis of posterior quantiles and HPPD regions. Several illustrative examples are considered in subsection 6.3.3.

6.3.1 Posterior predictive density

In the setup of Section 6.2, let $X_i = (W_i, Y_i)$ for each i, where W_i corresponds to an independent variable W and Y_i corresponds to a dependent variable Y, both W and Y being possibly vector-valued. Accordingly, $f(x; \theta)$ is rewritten as $f(w, y; \theta)$ which represents the joint density of (W, Y), or equivalently, the common joint density of the i.i.d. pairs (W_i, Y_i), $i \geq 1$. Clearly,

$$f(w, y; \theta) = f^{(1)}(w; \theta) f^{(2)}(y|w; \theta) ,\qquad (6.3.1)$$

where $f^{(1)}(w; \theta)$ is the marginal density of W and $f^{(2)}(y|w; \theta)$ is the conditional density of Y given $W = w$. Let $f_j^{(1)}(w; \theta) = D_j f^{(1)}(w; \theta)$ and $f_j^{(2)}(y|w; \theta) = D_j f^{(2)}(y|w; \theta)$.

As before, for $1 \leq i \leq n$, we regard $X_i = (W_i, Y_i)$ as a past observation where W_i and Y_i are both known. However, unlike in the last section, $X_{n+1} = (W_{n+1}, Y_{n+1})$ is now partially known in the sense that knowledge on W_{n+1} is available. We consider Bayesian prediction of the as yet unobserved Y_{n+1} on the basis of

$$Z = \{X_1, \ldots, X_n, W_{n+1}\} = \{(W_1, Y_1), \ldots, (W_n, Y_n), W_{n+1}\} ,\qquad (6.3.2)$$

using a prior density $\pi(\cdot)$.

By (6.2.4) and (6.3.1), the posterior density of (W_{n+1}, Y_{n+1}) $(= X_{n+1})$ given $X = (X_1, \ldots, X_n)^T$, under $\pi(\cdot)$, is represented by

$$\widetilde{\pi}(w_{n+1}, y_{n+1}|X)$$
$$= f(w_{n+1}, y_{n+1}; \widehat{\theta})$$
$$+ n^{-1}\{f_j^{(1)}(w_{n+1}; \widehat{\theta}) f^{(2)}(y_{n+1}|w_{n+1}; \widehat{\theta}) + f^{(1)}(w_{n+1}; \widehat{\theta}) f_j^{(2)}(y_{n+1}|w_{n+1}; \widehat{\theta})\}$$
$$\times c^{jr} \frac{\widetilde{\pi}_r}{\widehat{\pi}} + n^{-1} b(w_{n+1}, y_{n+1}, X) + o(n^{-1}) .\qquad (6.3.3)$$

Recall that $b(w_{n+1}, y_{n+1}, X)$, as given by (6.2.5), does not depend on $\pi(\cdot)$. Since

$$\int_{-\infty}^{\infty} f_j^{(2)}(y|w; \theta) dy = 0 ,$$

integrating (6.3.3) with respect to y_{n+1}, the posterior density of W_{n+1} given X, under $\pi(\cdot)$, equals

$$\tilde{\pi}(w_{n+1}|X) = f^{(1)}(w_{n+1};\hat{\theta}) + n^{-1}f_j^{(1)}(w_{n+1};\hat{\theta})c^{jr}\frac{\widehat{\pi_r}}{\widehat{\pi}} + n^{-1}b^*(w_{n+1}, X)$$
$$+ o(n^{-1}),\qquad\qquad (6.3.4)$$

where

$$b^*(w_{n+1}, X) = \int_{-\infty}^{\infty} b(w_{n+1}, y_{n+1}, X)dy_{n+1},$$

and the notation $\tilde{\pi}$ is used generically for any posterior density under $\pi(\cdot)$. From (6.3.1)–(6.3.4), it can be seen that the posterior predictive density of Y_{n+1} given Z, under $\pi(\cdot)$, admits an expansion of the form

$$\tilde{\pi}(y_{n+1}|Z) = f^{(2)}(y_{n+1}|W_{n+1};\hat{\theta}) + n^{-1}f_j^{(2)}(y_{n+1}|W_{n+1};\hat{\theta})c^{jr}\frac{\widehat{\pi_r}}{\widehat{\pi}}$$
$$+ n^{-1}\bar{b}(y_{n+1}, Z) + o(n^{-1}),\qquad\qquad (6.3.5)$$

where $\bar{b}(y_{n+1}, Z)$ does not depend on $\pi(\cdot)$ and is at most of order $O(1)$ for any fixed y_{n+1} (free from n). The detailed form of $\bar{b}(y_{n+1}, Z)$ is not required in the sequel. Observe that (6.3.5) is quite similar to (6.2.4).

6.3.2 Matching conditions

As in subsection 6.2.2, first suppose the dependent variable Y is scalar-valued so that posterior quantiles of Y_{n+1}, with reference to the predictive density (6.3.5), are well-defined. We assume that for each θ and w, the set $\{y : f^{(2)}(y|w;\theta) > 0\}$ is an interval, possibly unbounded, on the real line. Then for each θ, α $(0 < \alpha < 1)$ and w, there exists a unique $q(\theta, \alpha, w)$ such that

$$\int_{q(\theta,\alpha,w)}^{\infty} f^{(2)}(y|w;\theta)dy = \alpha.\qquad\qquad (6.3.6)$$

By (6.3.5) and (6.3.6), analogously to (6.2.8), the $(1-\alpha)$th posterior quantile of Y_{n+1}, given Z and under the prior $\pi(\cdot)$, is of the form

$$Q(\pi, Z, \alpha) = q(\hat{\theta}, \alpha, W_{n+1}) + O(n^{-1}).\qquad\qquad (6.3.7)$$

In consideration of (6.3.5) and (6.3.7), steps similar to those leading to (6.2.11) yield

$$P_\theta\{Y_{n+1} > Q(\pi, Z, \alpha)\} = \alpha - \frac{1}{n\pi(\theta)}D_r\{I^{jr}V_j(\theta, \alpha)\pi(\theta)\} + o(n^{-1}),$$

for all θ, where

$$V_j(\theta, \alpha) = E_\theta\{\int_{q(\theta,\alpha,W)}^{\infty} f_j^{(2)}(y|W;\theta)dy\}.\qquad\qquad (6.3.8)$$

Hence $\pi(\cdot)$ is a matching prior in the sense of ensuring frequentist validity, up to $o(n^{-1})$, of the posterior quantiles of Y_{n+1} if and only if it satisfies the partial differential equation

$$D_r\{I^{jr}V_j(\theta,\alpha)\pi(\theta)\} = 0 \qquad (6.3.9)$$

for every α.

We now turn to the situation where the dependent variable Y is possibly vector-valued. The HPPD regions for Y_{n+1} are still meaningful. For $0 < \alpha < 1$, let $m(\theta,\alpha,w)$ be such that $\int f^{(2)}(y|w;\theta)dy = 1-\alpha$, where the integral is over $A(\theta,\alpha,w) = \{y : f^{(2)}(y|w;\theta) \geq m(\theta,\alpha,w)\}$. Let $U_j(\theta,\alpha) = E_\theta\{\int f_j^{(2)}(y|W;\theta)dy\}$, the integral being over $A(\theta,\alpha,W)$. Then arguing as before, it can be shown that a prior $\pi(\cdot)$ ensures frequentist validity, up to $o(n^{-1})$, of HPPD regions for Y_{n+1} if and only if it satisfies the partial differential equation

$$D_r\{I^{jr}U_j(\theta,\alpha)\pi(\theta)\} = 0 \qquad (6.3.10)$$

for every α. The details are omitted to avoid repetition.

The same reasoning as in subsection 6.2.2 shows that the matching conditions (6.3.9) and (6.3.10) are invariant of the parameterization adopted.

6.3.3 Examples: applications to regression models

Example 6.3.1. With scalar Y and possibly vector-valued W, let

$$\left.\begin{aligned}
f^{(1)}(w;\theta) &= f^{(1)}(w;\theta_{(1)}) ,\\
f^{(2)}(y|w;\theta) &= \exp\{-\theta_{(2)}^T\tau(w)\}g[y\exp\{-\theta_{(2)}^T\tau(w)\}] ,
\end{aligned}\right\} \qquad (6.3.11)$$

where $\theta = (\theta_{(1)}^T, \theta_{(2)}^T)^T$. Here $g(\cdot)$ is a density on the real line and each of $\theta_{(1)}$, $\theta_{(2)}$ and $\tau(w)$ is possibly vector-valued, the functional form of $\tau(w)$ being known. Note that the independent variable W enters into the conditional distribution of Y via a scale parameter $\exp\{\theta_{(2)}^T\tau(w)\}$. The exponential regression model (Cox and Reid, 1987) is covered by (6.3.11) if one takes $g(v) = \exp(-v)$ for $v > 0$ and $= 0$ otherwise. Similarly, (6.3.11) covers the normal regression model with known coefficient of variation if one takes $g(v) = \phi(v - k)$, where $\phi(\cdot)$ is the standard univariate normal density and $k^{-1}(> 0)$ is the known coefficient of variation in the conditional distribution of Y.

In view of (6.3.6) and (6.3.8), after some algebra it can be checked that the following hold under (6.3.11): (a) I does not involve $\theta_{(2)}$; (b) $I = \text{diag}(M_1, M_2)$, where M_1 and M_2 correspond to $\theta_{(1)}$ and $\theta_{(2)}$ respectively; (c) $q(\theta,\alpha,w) = q_\alpha \exp\{\theta_{(2)}^T\tau(w)\}$, where q_α is the $(1-\alpha)$th quantile of the density $g(\cdot)$; (d) $V_j(\theta,\alpha)$ does not involve $\theta_{(2)}$ for any j; (e) $V_j(\theta,\alpha) = 0$ whenever j corresponds to some element of $\theta_{(1)}$. Hence one can verify that any prior $\pi(\cdot)$ that does not involve $\theta_{(2)}$ will satisfy the matching condition (6.3.9) for posterior quantiles. Furthermore, with additional algebra, the same conclusion is seen to hold also with respect to the matching condition (6.3.10) arising via HPPD regions. ♣

Example 6.3.2. Continuing with scalar Y and possibly vector-valued W now let

$$f^{(1)}(w;\theta) = f^{(1)}(w;\theta_{(1)}), \ f^{(2)}(y|w;\theta) = \delta^{-1}g[\{y - \theta_{(2)}^T\tau(w)\}/\delta] , \quad (6.3.12)$$

where $\theta = (\theta_{(1)}^T, \theta_{(2)}^T, \delta)^T$. As before $g(\cdot)$ is a density on the real line and each of $\theta_{(1)}$, $\theta_{(2)}$ and $\tau(w)$ is possibly vector-valued, the functional form of $\tau(w)$ being known. The independent variable W enters into the conditional distribution of Y via a location parameter $\theta_{(2)}^T\tau(w)$. Here $\delta(> 0)$ is a scale parameter underlying this conditional distribution. In particular, if W and Y are jointly normal (with no supplementary information available about the underlying parameters) then (6.3.12) arises and δ represents the conditional standard deviation of Y.

By (6.3.6) and (6.3.8), after some algebra one can check that the following hold under (6.3.12): (a) $I = \text{diag}(M_1, \delta^{-2}M_2)$, where M_1 and M_2 correspond to $\theta_{(1)}$ and $(\theta_{(2)}^T, \delta)^T$ respectively; (b) neither M_1 nor M_2 involves $\theta_{(2)}$ or δ; (c) $q(\theta, \alpha, w) = \delta q_\alpha + \theta_{(2)}^T\tau(w)$, where q_α is the $(1 - \alpha)$th quantile of the density $g(\cdot)$; (d) $V_j(\theta, \alpha)$ is of the form $V_j(\theta, \alpha) = \delta^{-1}G_j(\theta_{(1)}, \alpha)$ whenever j corresponds to δ or some element of $\theta_{(2)}$, $G_j(\theta_{(1)}, \alpha)$ being free from δ or $\theta_{(2)}$; (e) $V_j(\theta, \alpha) = 0$ whenever j corresponds to some element of $\theta_{(1)}$. Hence it may be verified that any prior $\pi(\cdot)$ of the form $\pi(\theta) = \kappa(\theta_{(1)})/\delta$, where $\kappa(\theta_{(1)})(> 0)$ is a smooth function involving $\theta_{(1)}$ alone, will satisfy the matching condition (6.3.9) for posterior quantiles. One can also check that the same conclusion holds for the matching condition (6.3.10) pertaining to HPPD regions.   ♣

Example 6.3.3. In the previous examples, the marginal density of W and the conditional density of Y given W involved no common parameter. Consider now a situation where this is not the case. Let the joint distribution of W and Y be bivariate normal with both means $\theta_1(\in \mathcal{R}^1)$, both standard deviations $\theta_2(> 0)$, and correlation coefficient θ_3 ($|\theta_3| < 1$). Then

$$I_{11} = 2/\{\theta_2^2(1 + \theta_3)\} , I_{12} = I_{13} = 0 , I_{22} = 4/\theta_2^2 ,$$

$$I_{23} = -2\theta_3/\{\theta_2(1 - \theta_3^2)\} , I_{33} = (1 + \theta_3^2)/(1 - \theta_3^2)^2 ,$$

$$V_1(\theta, \alpha) = \{(1 - \theta_3)/(1 + \theta_3)\}^{1/2}\phi(z_\alpha)/\theta_2 , V_2(\theta, \alpha) = z_\alpha\phi(z_\alpha)/\theta_2 ,$$

$$V_3(\theta, \alpha) = -\theta_3 z_\alpha\phi(z_\alpha)/(1 - \theta_3^2) ,$$

where, as before, z_α is the $(1 - \alpha)$th quantile of a standard normal variate and $\phi(\cdot)$ is the standard normal density. Hence considering a natural class of priors of the form $\pi(\theta) = \{\theta_2^r(1 - \theta_3^2)^s\}^{-1}$, where r and s are any real numbers, one can verify that the unique prior in this class satisfying the matching condition (6.3.9) is given by $r = -1$, $s = 1$. With additional algebra, the same conclusion is seen to hold for the matching condition (6.3.10). Thus, either via posterior quantiles or via HPPD regions, one gets the unique probability matching prior $\pi(\theta) = \theta_2/(1 - \theta_3^2)$ within the natural class mentioned above. Interestingly, in contrast with what was seen in Example 6.3.2, this prior is not inversely proportional to the conditional standard deviation of Y.   ♣

One can check that the matching prior obtained in the last example entails the propriety of the posterior predictive density, with P_θ−probability unity for all θ, whenever n is sufficiently large. Under wide generality, the same holds in Examples 6.3.1 and 6.3.2 as well.

Chang, Kim and Mukerjee (2003) reported results, formally similar to those of this section, on matching priors for predicting unobservable random effects. These results generalize an earlier one due to Datta, Ghosh and Mukerjee (2000). We refer to the original sources for details.

6.4 Concluding remarks

In this monograph, we reviewed probability matching priors for both estimation and prediction. This was done under the framework of i.i.d. absolutely continuous observations and standard regularity conditions. The sample size n was throughout supposed to be non-stochastic. There is considerable scope for further research on at least two of the topics discussed here. The first of these concerns subsection 2.8.2 on matching priors for one-sided Bayesian tolerance limits for a scalar-valued random variable. The corresponding results in the two-sided case or with a vector-valued random variable are as yet unknown. The second open issue relates to subsection 6.2.4 on prediction intervals, with both Bayesian and frequentist validity, for a scalar-valued future observation when the underlying parameter is also scalar-valued. More work is needed on the extension of this idea to situations where the future observation or the parameter is vector-valued.

Further challenging problems arise as soon as one steps beyond the framework considered here. In Sections 2.4 and 4.4, we referred to Ghosal (1999) and Rousseau (2000, 2002) for results in nonregular cases and discrete settings respectively. Evidently, there is much scope for work in these directions. Additional problems emerge when one attempts to consider a non-i.i.d. setup.

Even with i.i.d. absolutely continuous observations and standard regularity conditions, problems of significant interest arise when the sample size is stochastic as happens, for example, with censoring or a stopping rule. Sweeting (2001) investigated the crucial role played by data-dependent matching priors in such situations. Additional references on priors of this kind include Wasserman (2000) and Fraser and Reid (2002), and a brief review is available in Reid, Mukerjee and Fraser (2003). Further development of this area will greatly enrich the subject considered in this monograph.

References

1. Aitchison, J. and Dunsmore, I.R. (1975). *Statistical Prediction Analysis*. Cambridge University Press, Cambridge.

2. Amari, S. (1985). *Differential-Geometrical Methods in Statistics*. Springer-Verlag, Berlin.

3. Bar-Lev, S.K. and Reiser, B. (1982). An exponential subfamily which admits UMPU tests based on a single test statistic. *Ann. Statist.*, **10**, 979–989.

4. Barndorff-Nielsen, O.E. and Blæsild, P. (1986). A note on the calculation of Bartlett adjustments. *J. Roy. Statist. Soc. B*, **48**, 353–358.

5. Barndorff-Nielsen, O.E. and Cox, D.R. (1996). Prediction and asymptotics. *Bernoulli*, **2**, 319–340.

6. Barndorff-Nielsen, O.E. and Hall, P. (1988). On the level-error after Bartlett adjustment of the likelihood ratio statistic. *Biometrika*, **75**, 374–378.

7. Bartlett, M.S. (1937). Properties of sufficiency and statistical tests. *Proc. Roy. Soc. London A*, **160**, 268–282.

8. Bartlett, M.S. (1953). Approximate confidence intervals. *Biometrika*, **40**, 12–19.

9. Berger, J.O. and Bernardo, J.M. (1989). Estimating a product of means: Bayesian analysis with reference priors. *J. Amer. Statist. Assoc.*, **84**, 200–207.

10. Berger, J.O. and Bernardo, J.M. (1992a). On the development of reference priors (with discussion). In *Bayesian Statistics 4*, Eds. J.M. Bernardo, J.O. Berger, A.P. Dawid, and A.F.M. Smith, pp. 35–60, Oxford University Press, London.

11. Berger, J.O. and Bernardo, J.M. (1992b). Reference priors in a variance components problem. In *Bayesian Analysis in Statistics and Econometrics*, Eds. P.K. Goel and N.S. Iyenger, pp. 177–194, Springer-Verlag, Berlin.

12. Berger, J.O., Philippe, A. and Robert, C. (1998). Estimation of quadratic functions: noninformative priors for noncentrality parameters. *Statist. Sinica*, **8**, 359–375.

13. Bernardo, J.M. (1979). Reference posterior distributions for Bayesian inference (with discussion). *J. Roy. Statist. Soc. B*, **41**, 113–147.

14. Bernardo, J. M. and Ramon, J. M. (1998). An introduction to Bayesian reference analysis: Inference on the ratio of multinomial parameters. *J. Roy. Statist. Soc. D*, **47**, 101–135.

15. Bernardo, J.M. and Smith, A.F.M. (1994). *Bayesian Theory*. John Wiley, New York.

16. Bhattacharya, R.N. and Ghosh, J.K. (1978). On the validity of the formal Edgeworth expansion. *Ann. Statist.*, **6**, 434–451.

17. Bickel, P.J. and Ghosh, J.K. (1990). A decomposition for the likelihood ratio statistic and the Bartlett correction – a Bayesian argument. *Ann. Statist.*, **18**, 1070–1090.

18. Box, G.E.P. and Tiao, G.C. (1973). *Bayesian Inference in Statistical Analysis*. John Wiley, New York.

19. Chandra, T.K. (1980). *Asymptotic Expansions and Deficiency*. Unpublished Ph.D. dissertation. Indian Statistical Institute, Calcutta.

20. Chandra, T.K. and Ghosh, J.K. (1979). Valid asymptotic expansions for the likelihood ratio statistic and other perturbed chi-square variables. *Sankhyā A*, **41**, 22–47.

21. Chang, I.H., Kim, B.H. and Mukerjee, R. (2003). Probability matching priors for predicting unobservable random effects with application to ANOVA models. *Statist. Probab. Lett.*, **62**, 223–228.

22. Chung, Y. and Dey, D.K. (1998). Bayesian approach to estimation of intraclass correlation using reference prior. *Comm. Statist. A – Theory and Methods*, **27**, 2241–2255.

23. Clarke, B. and Wasserman, L. (1995). Information tradeoff. *Test*, **4**, 19–38.

24. Corcuera, J.M. and Giummole, F. (1999). A generalized Bayes rule for prediction. *Scand. J. Statist.*, **26**, 265–279.

25. Cox, D.R. and Reid, N. (1987). Parameter orthogonality and approximate conditional inference (with discussion). *J. Roy. Statist. Soc. B*, **49**, 1–39.

26. Cox, D.R. and Reid, N. (1993). A note on the calculation of adjusted profile likelihood. *J. Roy. Statist. Soc. B*, **55**, 467–471.

27. Creasy, M.A. (1954). Limits for the ratio of means. *J. Roy. Statist. Soc. B*, **16**, 186–194.

28. Datta, G.S. (1996). On priors providing frequentist validity for Bayesian inference for multiple parametric functions. *Biometrika*, **83**, 287–298.

29. Datta, G.S. and DiCiccio, T.J. (2001). On expected volumes of multidimensional confidence sets associated with the usual and adjusted likelihoods. *J. Roy. Statist. Soc. B*, **63**, 691–703.

30. Datta, G.S. and Ghosh, J.K. (1995a). Noninformative priors for maximal invariant parameter in group models. *Test*, **4**, 95–114.

31. Datta, G.S. and Ghosh, J.K. (1995b). On priors providing frequentist validity for Bayesian inference. *Biometrika*, **82**, 37–45.

32. Datta, G.S. and Ghosh, M. (1995a). Some remarks on noninformative priors. *J. Amer. Statist. Assoc.*, **90**, 1357–1363.

33. Datta, G.S. and Ghosh, M. (1995b). Hierarchical Bayes estimators of the error variance in one-way ANOVA models. *J. Statist. Plann. Inf.*, **45**, 399–411.

34. Datta, G.S. and Ghosh, M. (1996). On the invariance of noninformative priors. *Ann. Statist.*, **24**, 141–159.

35. Datta, G.S., Ghosh, M. and Kim, Y.H. (2002). Probability matching priors for one-way unbalanced random effects models. *Statist. Dec.*, **20**, 29–51.

36. Datta, G.S., Ghosh, M. and Mukerjee, R. (2000). Some new results on probability matching priors. *Calcutta Statist. Assoc. Bull.*, **50**, 179–192.

37. Datta, G.S. and Mukerjee, R. (2003). Probability matching priors for predicting a dependent variable with application to regression models. *Ann. Inst. Statist. Math.*, **55**, 1–6.

38. Datta, G.S., Mukerjee, R., Ghosh, M. and Sweeting, T.J. (2000). Bayesian prediction with approximate frequentist validity. *Ann. Statist.*, **28**, 1414–1426.

39. Dawid, A.P. (1991). Fisherian inference in likelihood and prequential frames of reference (with discussion). *J. Roy. Statist. Soc. B*, **53**, 79–109.

40. DiCiccio, T.J. and Stern, S.E. (1993). On Bartlett adjustments for approximate Bayesian inference. *Biometrika*, **80**, 731–740.

41. DiCiccio, T.J. and Stern, S.E. (1994). Frequentist and Bayesian Bartlett correction of test statistics based on adjusted profile likelihoods. *J. Roy. Statist. Soc. B*, **56**, 397–408.

42. Eno, D.R. and Ye, K. (2001). Probability matching priors for an extended statistical calibration model. *Can. J. Statist.*, **29**, 19–35.

43. Fieller, E.C. (1954). Some problems in interval estimation. *J. Roy. Statist. Soc. B*, **16**, 175–185.

44. Fraser, D.A.S. and Reid, N. (2002). Strong matching of frequentist and Bayesian parametric inference. *J. Statist. Plann. Inf.*, **103**, 263–285.

45. Garvan, C.W. and Ghosh, M. (1997). Noninformative priors for dispersion models. *Biometrika*, **84**, 976–982.

46. Garvan, C.W. and Ghosh, M. (1999). On the property of posteriors for dispersion models. *J. Statist. Plann. Inf.*, **78**, 229–241.

47. Geisser, S. (1965). Bayesian estimation in multivariate analysis. *Ann. Math. Statist.*, **36**, 150–159.

48. Geisser, S. (1993). *Predictive Inference: An Introduction*. Chapman and Hall, New York.

49. Ghosal, S. (1999). Probability matching priors for non-regular cases. *Biometrika*, **86**, 956–964.

50. Ghosh, J.K. (1994). *Higher Order Asymptotics*. Institute of Mathematical Statistics and American Statistical Association, Hayward, California.

51. Ghosh, J.K. and Mukerjee, R. (1991). Characterization of priors under which Bayesian and frequentist Bartlett corrections are equivalent in the multiparameter case. *J. Mult. Anal.*, **38**, 385–393.

52. Ghosh, J.K. and Mukerjee, R. (1992a). Non-informative priors (with discussion). In *Bayesian Statistics 4*, Eds. J.M. Bernardo, J.O. Berger, A.P. Dawid and A.F.M. Smith, pp. 195–210, Oxford University Press, London.

53. Ghosh, J.K. and Mukerjee, R. (1992b). Bayesian and frequentist Bartlett corrections for likelihood ratio and conditional likelihood ratio tests. *J. Roy. Statist. Soc. B*, **54**, 867–875.

54. Ghosh, J.K. and Mukerjee, R. (1993a). On priors that match posterior and frequentist distribution functions. *Can. J. Statist.*, **21**, 89–96.

55. Ghosh, J.K. and Mukerjee, R. (1993b). Frequentist validity of highest posterior density regions in multiparameter case. *Ann. Inst. Statist. Math.*, **45**, 293–302.

56. Ghosh, J.K. and Mukerjee, R. (1994a). Higher order comparison of tests: A Bayesian route. In *Essays on Probability and Statistics in Honor of Professor A.K. Bhattacharyya*, Eds. S.P. Mukherjee, A. Chaudhuri and S.K. Basu, pp. 92–101, Presidency College, Calcutta.

57. Ghosh, J.K. and Mukerjee, R. (1994b). Adjusted versus conditional likelihood: Power properties and Bartlett-type adjustment. *J. Roy. Statist. Soc. B*, **56**, 185–188.

58. Ghosh, J.K. and Mukerjee, R. (1995a). Frequentist validity of highest posterior density regions in the presence of nuisance parameters. *Statist. Dec.*, **13**, 131–139.

59. Ghosh, J.K. and Mukerjee, R. (1995b). On perturbed ellipsoidal and highest posterior density regions with approximate frequentist validity. *J. Roy. Statist. Soc. B*, **57**, 761–769.

60. Ghosh, J.K., Sinha, B.K. and Joshi, S.N. (1982). Expansions for posterior probability and integrated Bayes risk. In *Statistical Decision Theory and Related Topics III*, Eds. S.S. Gupta and J.O. Berger, pp. 403–456, Academic Press, New York.

61. Ghosh, M., Carlin, B.P. and Srivastava, M.S. (1995). Probability matching priors for linear calibration. *Test*, **4**, 333–357.

62. Ghosh, M. and Heo, J. (2003). Noninformative priors, credible sets and Bayesian hypothesis testing for the intraclass model. *J. Statist. Plann. Inf.*, **112**, 133–146.

63. Ghosh, M. and Kim, Y.-H. (2001). The Behrens-Fisher problem revisited: A Bayes-frequentist synthesis. *Can. J. Statist.*, **29**, 5–17.

64. Ghosh, M. and Mukerjee, R. (1998). Recent developments on probability matching priors. In *Applied Statistical Science III*, Eds. S.E. Ahmed, M. Ahsanullah and B.K. Sinha, pp. 227–252, Nova Science Publishers, New York.

65. Ghosh, M., Rousseau, J. and Kim, D. (2001). Noninformative priors for the bivariate Fieller-Creasy problem. *Statist. Dec.*, **19**, 277–288.

66. Ghosh, M. and Yang, M.C. (1996). Noninformative priors for the two-sample normal problem. *Test*, **5**, 145–157.

67. Ghosh M., Yin, M. and Kim, Y.H. (2003). Objective Bayesian inference for ratios of regression coefficients in linear models. *Statist. Sinica*, **13**, 409–422.

68. Guttman, I. (1970). *Statistical Tolerance Regions: Classical and Bayesian*. Charles Griffin, London.

69. Hora, R.B. and Buehler, R.J. (1966). Fiducial theory and invariant estimation. *Ann. Math. Statist.*, **37**, 643–656.

70. Huzurbazar, V.S. (1950). Probability distributions and orthogonal parameters. *Proc. Camb. Phil. Soc.*, **46**, 281–284.

71. Jeffreys, H. (1961). *Theory of Probability*. Oxford University Press, London.

72. Johnson, R.A. (1970). Asymptotic expansions associated with posterior distributions. *Ann. Math. Statist.*, **41**, 851–864.

73. Jorgensen, B. (1992). *The Theory of Exponential Dispersion Models and Analysis of Deviance*. Instituto de Matematica Pura e Aplicada: Rio de Janeiro.

74. Jorgensen, B. (1997). *The Theory of Dispersion Models*. Chapman and Hall, New York.

75. Kass, R.E. and Wasserman, L. (1996). The selection of prior distributions by formal rules. *J. Amer. Statist. Assoc.*, **91**, 1343–1370.

76. Kim, B.H., Chang, I.H. and Kang, C.K. (2001). Bayesian estimation for the reliability in Weibull stress-strength systems using noninformative priors. *Far East J. Theor. Statist.*, **5**, 299–315.

77. Kim, D.H., Kang, S.G. and Lee, W.D. (2001). Noninformative priors for intraclass correlation coefficient in familial data. *Far East J. Theor. Statist.*, **5**, 51–65.

78. Kim, D.H., Kang, S.G. and Lee, W.D. (2003). Noninformative priors for the nested design. Preprint.

79. Komaki, F. (1996). On asymptotic properties of predictive distributions. *Biometrika*, **83**, 299–313.

80. Kuboki, H. (1998). Reference priors for prediction. *J. Statist. Plann. Inf.*, **69**, 295–317.

81. Lee, C.B. (1989). *Comparison of Frequentist Coverage Probability and Bayesian Posterior Coverage Probability, and Applications*. Unpublished Ph.D. dissertation, Purdue University, Indiana.

82. Lee, G. (1998). Development of matching priors for $P(X < Y)$ in exponential distributions. *J. Korean Statist. Soc.*, **27**, 421–433.

83. Lehmann, E.L. (1986). *Testing Statistical Hypotheses* (2nd ed.) John Wiley, New York.

84. Levine, R.A. and Casella, G. (2003). Implementing matching priors for frequentist inference. *Biometrika*, **90**, 127–137.

85. Li, J. (1998). *Some Aspects of Bayesian and Frequentist Asymptotics*. Unpublished Ph.D. dissertation, University of Toronto, Ontario.

86. Lindley, D.V. (1958). Fiducial distributions and Bayes' theorem. *J. Roy. Statist. Soc. B*, **20**, 102–107.

87. Liseo, B.(1993). Elimination of nuisance parameters with reference priors. *Biometrika*, **80**, 295–304.

88. McCullagh, P. (1987). *Tensor Methods in Statistics*. Chapman and Hall, New York.

89. McCullagh, P. and Tibshirani, R. (1990). A simple method for the adjustment of profile likelihoods. *J. Roy. Statist. Soc. B*, **52**, 325–344.

90. Mukerjee, R. and Dey, D.K. (1993). Frequentist validity of posterior quantiles in the presence of a nuisance parameter: Higher order asymptotics. *Biometrika*, **80**, 499–505.

91. Mukerjee, R. and Ghosh, M. (1997). Second-order probability matching priors. *Biometrika*, **84**, 970–975.

92. Mukerjee, R. and Reid, N. (1999a). On a property of probability matching priors: Matching the alternative coverage probabilities. *Biometrika*, **86**, 333–340.

93. Mukerjee, R. and Reid, N. (1999b). On confidence intervals associated with the usual and adjusted likelihoods. *J. Roy. Statist. Soc. B*, **61**, 945–953.

94. Mukerjee, R. and Reid, N. (2000). On the Bayesian approach for frequentist computations. *Brazilian J. Probab. Statist.*, **14**, 159–166.

95. Mukerjee, R. and Reid, N. (2001). Second-order probability matching priors for a parametric function with application to Bayesian tolerance limits. *Biometrika*, **88**, 587–592.

96. Nicolaou, A. (1993). Bayesian intervals with good frequentist behaviour in the presence of nuisance parameters. *J. Roy. Statist. Soc. B*, **55**, 377–390.

97. Peers, H.W. (1965). On confidence sets and Bayesian probability points in the case of several parameters. *J. Roy. Statist. Soc. B*, **27**, 9–16.

98. Peers, H.W. (1968). Confidence properties of Bayesian interval estimates. *J. Roy. Statist. Soc. B*, **30**, 535–544.

99. Rao, C.R. (1973). *Linear Statistical Inference and Its Applications* (2nd ed.) John Wiley, New York.

100. Rao, C.R. and Mukerjee, R. (1995). On posterior credible sets based on the score statistic. *Statist. Sinica*, **5**, 781–791.

101. Reid, N. (1995). Likelihood and Bayesian approximation methods (with discussion). In *Bayesian Statistics 5*, Eds. J.M. Bernardo, J.O. Berger, A.P. Dawid and A.F.M. Smith, pp. 351–368, Oxford University Press, London.

102. Reid, N. (2003). Uniqueness of probability matching priors. Preprint.

103. Reid, N., Mukerjee, R. and Fraser, D.A.S. (2003) Some aspects of matching priors. In *Mathematical Statistics and Applications: Festschrift for Constance van Eeden*, Eds. M. Moore, S. Froda and C. Leger, pp. 31–44, CRM, Montreal.

104. Rousseau, J. (2000). Coverage properties of one-sided intervals in the discrete case and application to matching priors. *Ann. Inst. Statist. Math.*, **52**, 28–42.

105. Rousseau, J. (2002). Asymptotic properties of HPD regions in the discrete case. *J. Mult. Anal.*, **83**, 1–21.

106. Severini, T.A. (1991). On the relationship between Bayesian and non-Bayesian interval estimates. *J. Roy. Statist. Soc. B*, **53**, 611–618.

107. Severini, T.A. (1993). Bayesian interval estimates which are also confidence intervals. *J. Roy. Statist. Soc. B*, **55**, 533–540.

108. Severini, T.A. (1994). Approximately Bayesian inference. *J. Amer. Statist. Assoc.*, **89**, 242–249.

109. Severini, T.A., Mukerjee, R. and Ghosh, M. (2002). On an exact probability matching property of right-invariant priors. *Biometrika*, **89**, 952–957.

110. Stein, C. (1985). On the coverage probability of confidence sets based on a prior distribution. In *Sequential Methods in Statistics, Banach Center Publications 16*, pp. 485–514, Polish Scientific Publishers, Warsaw.

111. Sun, D. (1997). A note on noninformative priors for Weibull distributions. *J. Statist. Plann. Inf.*, **61**, 319–338.

112. Sun, D., Ghosh, M. and Basu, A.P. (1998). Bayesian analysis for a stress-strength system under noninformative priors. *Can. J. Statist.*, **26**, 323–332.

113. Sun, D. and Ye, K. (1995). Reference prior Bayesian analysis for normal mean products. *J. Amer. Statist. Assoc.*, **90**, 589–597.

114. Sun, D. and Ye, K. (1996). Frequentist validity of posterior quantiles for a two-parameter exponential family. *Biometrika*, **83**, 55–65.

115. Sun, D. and Ye, K. (1999). Reference priors for a product of normal means when variances are unknown. *Can. J. Statist.*, **27**, 97–103.

116. Sweeting, T.J. (1999). On the construction of Bayes-confidence regions. *J. Roy. Statist. Soc. B*, **61**, 849–861.

117. Sweeting, T.J. (2001). Coverage probability bias, objective Bayes and the likelihood principle. *Biometrika*, **88**, 657–675.

118. Sweeting, T.J., Datta, G.S. and Ghosh, M. (2003). Nonsubjective priors via predictive entropy loss. Preprint.

119. Tibshirani, R.J. (1989). Noninformative priors for one parameter of many. *Biometrika*, **76**, 604–608.

120. Vidoni, P.A. (1998). A note on modified estimative prediction limits and distributions. *Biometrika*, **85**, 949–953.

121. Wang, C-P. and Ghosh. M. (2000). Bayesian analysis of bivariate competing risks models. *Sankhyā B*, **62**, 388–401.

122. Wasserman, L. (2000). Asymptotic inference for mixture models using data-dependent priors. *J. Roy. Statist. Soc. B*, **62**, 159–180.

123. Welch, B.L. and Peers, H.W. (1963). On formulae for confidence points based on integrals of weighted likelihoods. *J. Roy. Statist. Soc. B*, **25**, 318–329.

124. Woodroofe, M. (1986). Very weak expansions for sequential confidence levels. *Ann. Statist.*, **14**, 1049–1067.

125. Yang, R. and Berger, J.O. (1997). A catalog of noninformative priors. ISDS Discussion Paper, Duke University, Durham, North Carolina.

124   References

126. Ye, K. (1994). Bayesian reference prior analysis on the ratio of variances for the balanced one-way random effect model. *J. Statist. Plann. Inf.*, **41**, 267–280.

127. Yin, M. (2000). Noninformative priors for multivariate linear calibration. *J. Mult. Anal.*, **73**, 221–240.

128. Yin, M. and Ghosh, M. (1997). A note on the probability difference between matching priors based on posterior quantiles and on inversion of conditional likelihood ratio statistics. *Calcutta Statist. Assoc. Bull.*, **47**, 59–65.

129. Yin, M. and Ghosh, M. (2001). Bayesian and likelihood inference for the generalized Fieller–Creasy problem. In *Empirical Bayes and Likelihood Inference*, Eds. S.E. Ahmed and N. Reid, pp. 121–139, Springer-Verlag, Berlin.

Index

Lecture Notes in Statistics

For information about Volumes 1 to 122, please contact Springer-Verlag

123: D. Y. Lin and T. R. Fleming (Editors), Proceedings of the First Seattle Symposium in Biostatistics: Survival Analysis. xiii, 308 pp., 1997.

124: Christine H. Müller, Robust Planning and Analysis of Experiments. x, 234 pp., 1997.

125: Valerii V. Fedorov and Peter Hackl, Model-Oriented Design of Experiments. viii, 117 pp., 1997.

126: Geert Verbeke and Geert Molenberghs, Linear Mixed Models in Practice: A SAS-Oriented Approach. xiii, 306 pp., 1997.

127: Harald Niederreiter, Peter Hellekalek, Gerhard Larcher, and Peter Zinterhof (Editors), Monte Carlo and Quasi-Monte Carlo Methods 1996. xii, 448 pp., 1997.

128: L. Accardi and C.C. Heyde (Editors), Probability Towards 2000. x, 356 pp., 1998.

129: Wolfgang Härdle, Gerard Kerkyacharian, Dominique Picard, and Alexander Tsybakov, Wavelets, Approximation, and Statistical Applications. xvi, 265 pp., 1998.

130: Bo-Cheng Wei, Exponential Family Nonlinear Models. ix, 240 pp., 1998.

131: Joel L. Horowitz, Semiparametric Methods in Econometrics. ix, 204 pp., 1998.

132: Douglas Nychka, Walter W. Piegorsch, and Lawrence H. Cox (Editors), Case Studies in Environmental Statistics. viii, 200 pp., 1998.

133: Dipak Dey, Peter Müller, and Debajyoti Sinha (Editors), Practical Nonparametric and Semiparametric Bayesian Statistics. xv, 408 pp., 1998.

134: Yu. A. Kutoyants, Statistical Inference For Spatial Poisson Processes. vii, 284 pp., 1998.

135: Christian P. Robert, Discretization and MCMC Convergence Assessment. x, 192 pp., 1998.

136: Gregory C. Reinsel, Raja P. Velu, Multivariate Reduced-Rank Regression. xiii, 272 pp., 1998.

137: V. Seshadri, The Inverse Gaussian Distribution: Statistical Theory and Applications. xii, 360 pp., 1998.

138: Peter Hellekalek and Gerhard Larcher (Editors), Random and Quasi-Random Point Sets. xi, 352 pp., 1998.

139: Roger B. Nelsen, An Introduction to Copulas. xi, 232 pp., 1999.

140: Constantine Gatsonis, Robert E. Kass, Bradley Carlin, Alicia Carriquiry, Andrew Gelman, Isabella Verdinelli, and Mike 142: György Terdik, Bilinear Stochastic Models and Related Problems of Nonlinear Time Series Analysis: A Frequency Domain West (Editors), Case Studies in Bayesian Statistics, Volume IV. xvi, 456 pp., 1999.

141: Peter Müller and Brani Vidakovic (Editors), Bayesian Inference in Wavelet Based Models. xiii, 394 pp., 1999.

142: György Terdik, Bilinear Stochastic Models and Related Problems of Nonlinear Time Series Analysis: A Frequent Domain Approach. xi, 258 pp., 1999.

143: Russell Barton, Graphical Methods for the Design of Experiments. x, 208 pp., 1999.

144: L. Mark Berliner, Douglas Nychka, and Timothy Hoar (Editors), Case Studies in Statistics and the Atmospheric Sciences. x, 208 pp., 2000.

145: James H. Matis and Thomas R. Kiffe, Stochastic Population Models. viii, 220 pp., 2000.

146: Wim Schoutens, Stochastic Processes and Orthogonal Polynomials. xiv, 163 pp., 2000.

147: Jürgen Franke, Wolfgang Härdle, and Gerhard Stahl, Measuring Risk in Complex Stochastic Systems. xvi, 272 pp., 2000.

148: S.E. Ahmed and Nancy Reid, Empirical Bayes and Likelihood Inference. x, 200 pp., 2000.

149: D. Bosq, Linear Processes in Function Spaces: Theory and Applications. xv, 296 pp., 2000.

150: Tadeusz Caliński and Sanpei Kageyama, Block Designs: A Randomization Approach, Volume I: Analysis. ix, 313 pp., 2000.

151: Håkan Andersson and Tom Britton, Stochastic Epidemic Models and Their Statistical Analysis. ix, 152 pp., 2000.
152: David Ríos Insua and Fabrizio Ruggeri, Robust Bayesian Analysis. xiii, 435 pp., 2000.

153: Parimal Mukhopadhyay, Topics in Survey Sampling. x, 303 pp., 2000.

154: Regina Kaiser and Agustín Maravall, Measuring Business Cycles in Economic Time Series. vi, 190 pp., 2000.

155: Leon Willenborg and Ton de Waal, Elements of Statistical Disclosure Control. xvii, 289 pp., 2000.

156: Gordon Willmot and X. Sheldon Lin, Lundberg Approximations for Compound Distributions with Insurance Applications. xi, 272 pp., 2000.

157: Anne Boomsma, Marijtje A.J. van Duijn, and Tom A.B. Snijders (Editors), Essays on Item Response Theory. xv, 448 pp., 2000.

158: Dominique Ladiray and Benoît Quenneville, Seasonal Adjustment with the X-11 Method. xxii, 220 pp., 2001.

159: Marc Moore (Editor), Spatial Statistics: Methodological Aspects and Some Applications. xvi, 282 pp., 2001.

160: Tomasz Rychlik, Projecting Statistical Functionals. viii, 184 pp., 2001.

161: Maarten Jansen, Noise Reduction by Wavelet Thresholding. xxii, 224 pp., 2001.

162: Constantine Gatsonis, Bradley Carlin, Alicia Carriquiry, Andrew Gelman, Robert E. Kass Isabella Verdinelli, and Mike West (Editors), Case Studies in Bayesian Statistics, Volume V. xiv, 448 pp., 2001.

163: Erkki P. Liski, Nripes K. Mandal, Kirti R. Shah, and Bikas K. Sinha, Topics in Optimal Design. xii, 164 pp., 2002.

164: Peter Goos, The Optimal Design of Blocked and Split-Plot Experiments. xiv, 244 pp., 2002.

165: Karl Mosler, Multivariate Dispersion, Central Regions and Depth: The Lift Zonoid Approach. xii, 280 pp., 2002.

166: Hira L. Koul, Weighted Empirical Processes in Dynamic Nonlinear Models, Second Edition. xiii, 425 pp., 2002.

167: Constantine Gatsonis, Alicia Carriquiry, Andrew Gelman, David Higdon, Robert E. Kass, Donna Pauler, and Isabella Verdinelli (Editors), Case Studies in Bayesian Statistics, Volume VI. xiv, 376 pp., 2002.

168: Susanne Rässler, Statistical Matching: A Frequentist Theory, Practical Applications, and Alternative Bayesian Approaches. xviii, 238 pp., 2002.

169: Yu. I. Ingster and Irina A. Suslina, Nonparametric Goodness-of-Fit Testing Under Gaussian Models. xiv, 453 pp., 2003.

170: Tadeusz Caliński and Sanpei Kageyama, Block Designs: A Randomization Approach, Volume II: Design. xii, 351 pp., 2003.

171: David D. Denison, Mark H. Hansen, Christopher C. Holmes, Bani Mallick, and Bin Yu (Editors), Nonlinear Estimation and Classification. viii, 474 pp., 2003.

172: Sneh Gulati and William J. Padgett, Parametric and Nonparametric Inference from Record-Breaking Data. viii, 111 pp., 2003.

173: Jesper Møller (Editor), Spatial Statistics and Computational Methods. xiv, 202 pp., 2003.

174: Yasuko Chikuse, Statistics on Special Manifolds. xxvi, 399 pp., 2003.

175: Jürgen Gross, Linear Regression. xiv, 394 pp., 2003.

176: Zehua Chen, Zhidong Bai, and Bimal K. Sinha, Ranked Set Sampling: Theory and Application. xii, 224 pp., 2004.

177: Caitlin Buck and Andrew Millard (Editors), Tools for Constructing Chronologies: Crossing Disciplinary Boundaries. xvi, 263 pp., 2004.

178: Gauri Sankar Datta and Rahul Mukerjee, Probability Matching Priors: Higher Order Asymptotics. x, 126 pp., 2004.

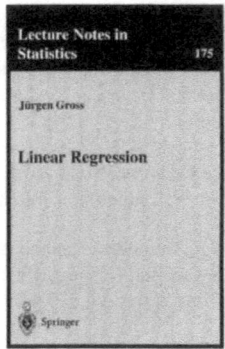